Appomattox

The Passing of the Armies

By

James W. Wensyel, Colonel, USA (Ret.)

 WHITE MANE BOOKS

This White Mane Books publication
was printed by
Beidel Printing House, Inc.
63 West Burd Street
Shippensburg, PA 17257-0152 USA

In respect for the scholarship contained herein, the acid-free paper used in this book meets the guidelines for permanence and durability of the Committee on Production Guidelines for Book Longevity of the Council on Library Resources.

For a complete list of available publications
please write
White Mane Books
Division of White Mane Publishing Company, Inc.
P.O. Box 152
Shippensburg, PA 17257-0152 USA

Library of Congress Cataloging-in-Publication Data

Wensyel, James W., 1928-
    Appomattox : the passing of the armies / by James W. Wensyel.
        p.   cm.
    Includes bibliographical references (p.   ).
    ISBN 1-57249-155-8 (acid-free paper)
    1. Appomattox Campaign, 1865.   I. Title.
E477.67.W46   1999
973.7'38--dc21   99-16698
                                                                      CIP

PRINTED IN THE UNITED STATES OF AMERICA

To "Sam," the citizen-soldier of the American Army, whose courage, endurance, and sense of humor have never failed him in more than 200 years of our country's freedom.

Contents

Illustrations

Foreword

Monday, April 3, 1865. In Virginia, two American armies have fought each other for nearly four years. Lieutenant General Ulysses S. Grant's Union army (comprised of Major General George Gordon Meade's Army of the Potomac, Major General Edward O. C. Ord's Army of the James, and Major General Philip H. Sheridan's cavalry); and Lieutenant General Robert E. Lee's Army of Northern Virginia (its two corps of infantry under Major General James Longstreet and Major General John B. Gordon, and its cavalry under Major General Fitzhugh Lee).

Magnificent armies with magnificent soldiers; mostly volunteers, bloodied veterans who've survived the near constant fighting. A great many of them still willing to die for causes they believe in: Billy Yank to preserve the Union and to end slavery; Johnny Reb to live his life without interference from a strong central government hundreds of miles away.

More than 600,000 Americans already have died in this civil war. Both sides are tired of it; and the country is financially, physically, and emotionally drained by it. Everyone is anxious for peace; but neither side is willing to compromise its cause.

For the past nine months, until two days ago, the fighting in the East has focused upon Grant's siege of Richmond, the Confederate capital; and Petersburg, two-dozen miles below Richmond and key to control of the Southside Railroad, Lee's rail link for men and supplies from the other Southern states. For those nine long months Lee's ragged, hungry soldiers (official rations less than 1,000 calories a day but that seldom available), spaced far apart in their trenches and with few reserves, have fought Grant's well-fed, well-equipped, three or four times their numbers troops to a standstill.

Then on April 1, Sheridan's cavalry, reinforced with Meade's V Corps of infantry under Major General Charles Griffin, turned Lee's right flank at Five Forks, a little woods' road junction southwest of Petersburg. After scattering Confederate Major General George E. Pickett's infantry and Fitz Lee's cavalry there, Sheridan marched north to cut the Southside Railroad. When Lee weakened the center of his line to counter Sheridan, the remainder of Meade's Army of the Potomac and Ord's Army of the James struck Lee's front, breaking it in several places. Lee, however, managed to hold open a slim corridor through Petersburg for his survivors to fall back across the Appomattox River, north of the city.

Lee's hope now is to withdraw to the west where he might reach the Richmond-Danville Railroad, be resupplied, then move south to unite with the only other remaining Confederate army in the East, that under Lieutenant General Joseph E. Johnston, opposing the northward advance of Union Major General William T. Sherman in North Carolina. Grant hopes to block Lee's move; to get ahead of him then force Lee's army to fight in the open or to surrender.

At City Point, below Petersburg, Grant's headquarters before he took to the field for this campaign, Union President Abraham Lincoln carefully plots Grant's reports on a worn campaign map and encourages his general to push hard to end the war without another bloody battle.

Confederate President Jefferson Davis, alerted by Lee that Petersburg has fallen and that Richmond also must be abandoned, has fled to the relative safety of Danville, Virginia with his cabinet, the Confederate treasury, and as much of its government as the remaining decrepit trains can carry.

Northern troops have entered Petersburg and Richmond, stacked their arms, and now work to put out widespread fires and restore order.

Below the Appomattox River, Sheridan's cavalry range ahead of Grant's infantry, marching hard to cut the Richmond-Danville Railroad and block Lee's retreat. Above the river, with a slight initial lead, Lee's army hurries toward tiny Amelia Court House where they've been promised that rations and ammunition will be waiting. They plan a brief rest there, then a forced march to regain their lead on Grant's pursuing armies.

Preface

The story of the fighting that finally breaks Confederate General Robert E. Lee's lines before Petersburg, Virginia and forces Lee to begin a desperate march to unite his army with that of Confederate General Joseph E. Johnston in North Carolina is told in the companion volume *PETERSBURG*.

APPOMATTOX picks up where *PETERSBURG* ends. It is the story of Lee's retreat and the pursuit by the Union armies of General Ulysses S. Grant; the running battles that occur when those armies contact each other; and the surrender of the Confederate army at Appomattox Court House, Virginia, on April 9, 1865.

APPOMATTOX is narrative history, told from the viewpoints of the presidents, generals, soldiers, and civilians who were there, individuals who with the exception of Sergeant Thomas McDermott, really existed. I'm sure that there was *a* "Sergeant McDermott" and that he was invaluable to Brigadier General Joshua Chamberlain; "Sergeant McDermotts" always are.

With that exception my story is historically correct; and I've included a selected bibliography of other authors who contributed to it.

I've condensed some of the action and given certain individuals thoughts and words that seemed reasonable to me and which, I believe, add to the overall picture. It's also likely that I've made their manner of speaking less formal than the words those individuals reported in their reminiscences. Men are inclined to be less formal on the battlefield than they will later recall. The character I assign to each of them reflects my own thoughts about that individual; you may agree or disagree as you come to know them better.

Our Civil War is a very important part of our national and personal heritage. The Petersburg and Appomattox Campaigns were a critical and poignant part of that war. Thank you for your interest. Enjoy!

Acknowledgments

This has been a work of love, and I'm very grateful for the help others have given me in completing it.

My thanks to Doctor Gerald A. Leidy for his photography; to Lieutenant David Weaver, Licensed Battlefield Guide at Gettysburg and cartographer of my story; to my perceptive editor, Harold Collier, who took the chance; to Chris M. Calkins, Park Ranger/Historian, Petersburg National Battlefield Historic Park, who encouraged my work and whose texts contributed so much to it; to Joe Williams, Curator, Appomattox Court House National Historical Park, who answered many questions; to Miss Susan Ravdin, Special Collections staff of the Bowdoin College Library, Brunswick, Maine, for her insights about General Joshua L. Chamberlain; to the Military History Institute, Carlisle Barracks, Pennsylvania, for texts and photographs; to Michael Moss, Director of the West Point Museum, U.S. Military Academy, for the print "The Surrender of the Army of Northern Virginia" by Ken Riley; to Burke Davis and all the other authors cited in the bibliography, who provided facts to be woven into the story; to Catherine E. Shirk, whose faith helped so much a long time ago; and to my wife, Jean, for her love and support from the beginning.

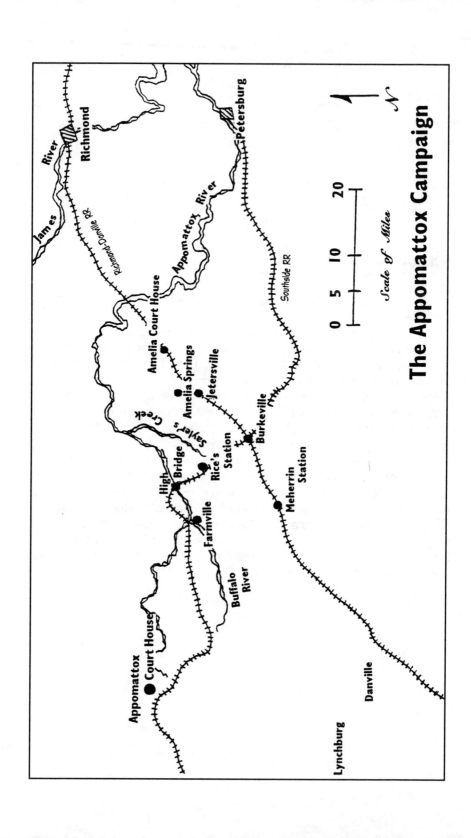

The Appomattox Campaign

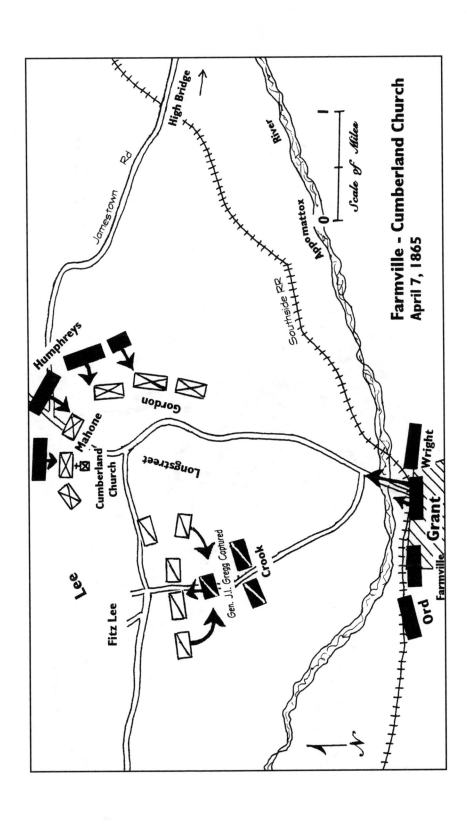

Farmville - Cumberland Church
April 7, 1865

Scale of Miles

High Bridge

Jamestown Rd

Appomattox River

Southside RR

Humphreys

Gordon

Mahone

Cumberland Church

Longstreet

Lee

Fitz Lee

Gen. J.I. Gregg Captured

Crook

Wright

Grant

Ord

Farmville

N

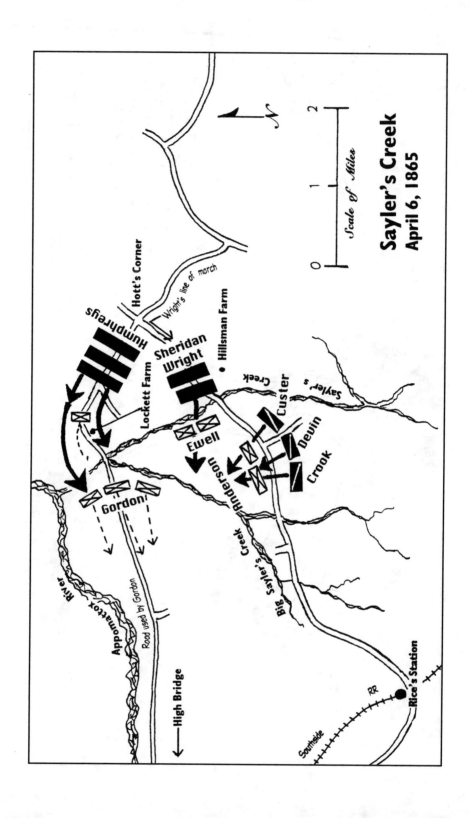

Sayler's Creek
April 6, 1865

Scale of Miles

0   1   2

N

Hott's Corner

Wright's line of march

Humphreys

Sheridan
Wright

Hillsman Farm

Lockett Farm

Sayler's Creek

Custer

Ewell

Devin

Anderson

Crook

Gordon

Sayler's Creek

Big Sayler's Creek

Appomattox River

Road used by Gordon

High Bridge

Southside

RR

Rice's Station

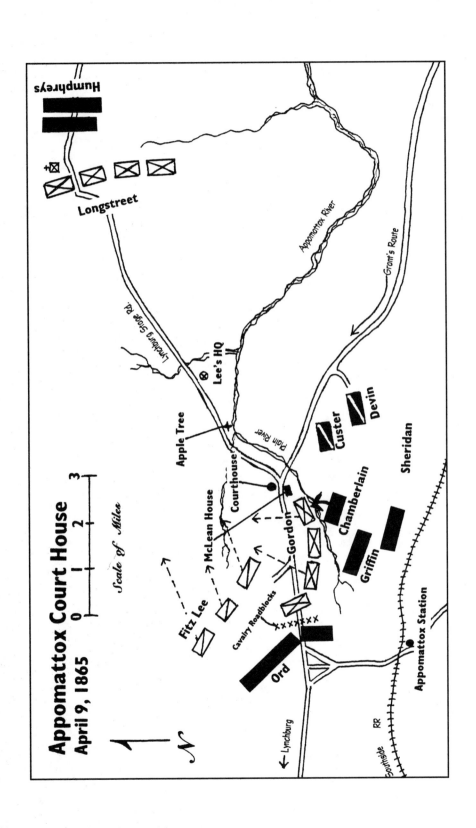

Appomattox Court House
April 9, 1865

Scale of Miles

0 1 2 3

N

Humphreys

Longstreet

Lynchburg Stage Rd.

Apple Tree

Lee's HQ

Appomattox River

Grant's Route

McLean House

Courthouse

Plain River

Custer

Devin

Fitz Lee

Gordon

Chamberlain

Sheridan

Griffin

Cavalry Roadblocks

Ord

Lynchburg

Appomattox Station

Southside RR

CHAPTER 1

The Long March   April 3, 1865

THE SOLDIERS

Perhaps 30,000 badly mauled soldiers of Lee's army struggle to reach Amelia Court House where they've been promised that food and ammunition are waiting. Most still carry their rifles, but others simply trudge along, heads down, without weapons. More a burden to Lee than a help. The heart of the Gray army, however, is not yet ready to admit defeat.

"Amelia Court House," the most stubborn insist, "there's hardtack there and cartridges. Then we'll march on down to Carolina and Joe Johnston's army. Whup Sherman; then Uncle Robert'll turn on Grant like a bear with a thorn in its paw and, afore you know it, we'll turn this war 'round again."

Confederate soldiers, so happy to be out of the cold, wet, miserable, deadly trenches to where they can maneuver as they did so well in the early days that, hungry or not, they begin to revive; most of them still confident that somewhere up ahead Lee will lead them to victory, as he's always done.

———

Grant's pursuing armies are strong: more than 100,000 well-fed, well-equipped veterans. If not rested then more rested than the Johnnies ahead of them. Told by their officers, "We've got 'em right where we want 'em, boys! We've got the twist on 'em! They can't turn back, and Sheridan's cavalry's up ahead drivin' hard to cut off their retreat. Cut 'em off; then we'll whip 'em once and for all. Your legs must do it boys; your legs must do it! The end of the war's in sight!"

Of course they'd heard *that* before, and a lot of Grant's soldiers don't believe the good news this time either.

———

About mid-morning, couriers gallop from regiment to regiment, spreading additional news: Richmond is burning, being evacuated.

They know about Petersburg; their fighting had seen to that. And the regiments that passed through the city tell of seeing Grant standing in a Petersburg doorway ignoring Southern snipers taking potshots at him. "Unconditional Surrender" Grant, General of all the Union armies, in his private's uniform giving orders and returning soldiers' cheers with a smile while his cigar waved them west. Yes, Petersburg's fallen.

But Richmond taken too? Richmond's been their goal, that dim light way out on the horizon they've been trying to reach for nearly four years. The Stars and Stripes flying over Richmond? Hard to believe. Why that would mean the war *has* to be nearly over. Hard to believe.

When their officers announce it, skeptical veterans remember how those same officers, just to make them step livelier, lied to them in the past. And when their company commanders add that the black soldiers of Major General Godfrey Weitzel's XXV Corps have actually entered Richmond, those skeptical veterans discreetly mumble among themselves, "Put it in a canteen!" "Tell it to the recruits!"

But when their colonels (there's something far more sacred about colonels than captains) confirm it, they finally believe it. Believe it and realize that this time they're being let in on the generals' strategy. Realize that this time that light they must reach somewhere down the road ahead of them is not some city or bitterly held earthwork but is the *road* itself. Cut that road, anywhere ahead of Lee's army, and Johnny'll have to surrender. Now *that*, they all understand, really *could* end this damned war, once and for all.

Understanding the reason for their forced march, they don't complain as much as they might; and that evening, over a short bivouac fire, one Union veteran shares with another, "Jack, I feel better tonight; even better than I did after that fight at Gettysburg."

They'd feel better yet if they could see Richmond and Petersburg burning behind them or Lee's army, also hurrying west, but not so fast, above the Appomattox River.

––––––––

In Richmond Union General Weitzel has ordered Colonel E. H. Ripley to put out the fires and restore order.

All day long, bolstered by Richmond's Mayor Joseph Mayo and all the volunteers and black laborers they can impress, Ripley's soldiers fight fires, round up left-behind Confederate soldiers, and lock arrested looters in Libby and Castle Thunder prisons' cells, cells that a few days before held Union prisoners of war.

One of Ripley's officers is fighting a fire in a Franklin Street church when a Negro calls to him from the door of a nearby brick home.

"Cap'n, suh, my mistress wants to speaks with you. In here, suh."

Entering, the officer meets a young woman who explains.

"Captain, I've watched you fighting the fires. Your soldiers seem well behaved, but some of our homes have been entered and searched by Yankee soldiers. My mother, because she's not well, can't leave our home, and she musn't be disturbed. Won't you post a guard?"

"I will, ma'am. Immediately. But you seem to know military terms; may I have your name?"

"Lee, Captain. My father is General Robert E. Lee; and two other members of our family also are generals in the Confederate Army. My father and mother know many of your generals from the Old Army. We'd be grateful for your courtesy and protection."

Within an hour a corporal and two soldiers of the 9th Vermont Infantry Regiment stand guard before the Lee home while an army ambulance is kept nearby.

In the afternoon a Negro reporter with the Philadelphia *Press*, J. Morris Chester, seats himself at the abandoned Speaker's table of the Confederate Congress and begins to write his impressions of the fallen city.

Then a paroled Confederate officer in uniform passes by, sees Chester sitting at the Speaker's table, and storms into the hall.

"Get out of that chair, damn you!"

Chester glances up then, dismissing the Southerner, resumes his writing.

Infuriated, the Confederate captain bellows again, "Come out of there or I'll knock your brains out!"

When Chester again ignores him, the Gray officer reaches down to grab the lapels of the reporter's coat.

Chester, however, merely shakes off his tormentor and again takes his seat.

A white Federal officer, hearing the noise, has come into the hall. Seeing him, the Confederate, his face beet-red, shouts.

"Captain, lend me your saber. I'll cut out this damned Nigger's heart!"

"No," the Union captain shakes his head. "But, if you want a fight, I'll see that no one interferes while the two of you go at it, man-to-man. You'll have no weapon or any other help though, and I'll bet that he'll give you the thrashing you need."

The Southerner, seeing Chester quietly standing and no sympathy from the Federal officer, hurries from the room.

When he's gone, Chester nods his thanks then quietly sits to take up his story where he'd been interrupted.

Although most of Richmond's women and children spend the day peeping at their Yankee conquerors from heavily curtained or shuttered windows, by late afternoon many of Richmond's men have returned to the streets and even begun to speak with Union soldiers. Soldiers who behave well. There is no more looting, no more fires, few confrontations between conqueror and conquered. And, except for the dozens of military bands that impudently play Union songs and "The Star Spangled Banner" again and again throughout the day, most Union soldiers seem to go out of their way to ease the citizens' concerns. Richmond will spend its first night back in the Federal Union quietly.

One Richmond citizen, Confederate nurse Phoebe Yates Pember, however, does not.

PEMBER

All morning, from Chimborazo Hospital, Phoebe Yates Pember watches Weitzel's soldiers streaming toward Richmond. Most of the regiments skirt the city, crossing the James River and marching west after Lee's retreating army. Only a few regiments, Ripley's men, enter the city itself. Some Union soldiers approach Chimborazo, but none reaches Phoebe Pember's wards until mid-afternoon.

She's been very busy with the Confederate wounded who were unable to march with Lee's army. Some are so badly wounded that they're quiet, resigned to their fate; others are anxious about the treatment they can expect from the Yankees. Pember has worked hard to reassure them, to feed and bathe them, and to treat their wounds, particularly difficult with only herself and the ward boy, Jim.

Now, in mid-afternoon, her work is interrupted again.

Glancing up from changing the dressing on an amputated leg, she realizes that the hospital's lone remaining surgeon is waiting at the door. Stifling a frown at the interruption, she straightens from her patient to smooth her apron and to coax a stubborn lock of hair back into place before turning to him.

Then she realizes that he's not alone. Also waiting at the door are three Union officers. Behind her the conversation among her patients stops; they've seen the Northern uniforms too. She turns to reassure them with a wave of her hands and a smile she hardly feels. Then she turns again to face her visitors.

"Yes, Doctor?"

"Mrs. Pember. Ma'am, I'm right sorry to interrupt your work. Permit me to introduce these gentlemen, all surgeons with General Grant's army."

Then he tells her, "We must ask you to care for a dozen more patients, Union soldiers. Tomorrow they'll be transferred to the City Point hospital, but for tonight they must remain here."

"Very well, Doctor," she lowers her voice, "but is it wise to mix patients this way? Our men have seen heavy fighting; so have theirs. Perhaps they should not be put together."

"Mrs. Pember," one of the Northern doctors smiles, "we appreciate your concern, but I think that you'll find that they'll get along fine. We've seen this happen in the past, on the battlefields. For these men, and I pray for all of us, this war is about over. They'll get along fine."

Not quite convinced but accepting the inevitable, the pretty nurse returns his smile, "Of course, Doctor."

Then she seizes the opportunity, "Now I'll need rations for your men and for my own. Whatever you can spare; as much as you can spare. Our diet has been rather restricted. We have enough to share with your patients tonight, but please have provisions delivered here early in the morning. And, if you can spare an orderly or two who won't mind working for a woman, please have them report to me as quickly as possible. I shall need all the help I can get."

"Of course, ma'am," the Federal surgeon smiles again. "You handle soldiers well; my men will be in good hands."

When the surgeons have gone, Pember, to whom the Richmond Commissary had issued no food before evacuating the hospital, turns to her patients, smiles, and tells them, "You see, I told you that the Lord would provide!"

And she needn't have worried about mixing Northern and Southern patients. Soon after the wounded Union soldiers have been settled on cots, Billy Yank and Johnny Reb are chatting comfortably; laughing; sharing tobacco; swapping pocket knives, Confederate money, and other trinkets.

They get along so well that Pember has trouble getting them to sleep before returning to her own small room nearby.

Suddenly awakened by a loud crash, she lights a lantern, wraps herself in a heavy robe, and opens her door to find a half dozen rough-looking men breaking into the ward's pantry.

Their ringleader, a dirty, unshaven, hard-faced soldier named Wilson who's given her much trouble in the past, apparently has deserted the Confederate army. His companions, she guesses, also are deserters. She knows several more of them as former patients she'd called "hospital rats" because her surgeon, finding their wounds or illnesses slight or imaginary, had ordered them back to their units only to have them return to Chimborazo in a week or so under a new guise. Malingerers. Wilson also had bullied patients and stolen their rations until she'd put a stop to it. She'd breathed a sigh of relief when they'd gone. Now they're back, as deserters, for whatever they can steal.

Realizing that she must look absurd, a hundred pounds of feminine authority and so small that she must hold her lantern high above her head to see Wilson's face, she gambles that in the shadows he can't see how frightened she is.

"Well, Wilson," with all the authority she can muster she asks, "what do you want?"

"We've come for the whiskey, Missy. The keg you keep back there. Don't need none of your sass, thank you."

Inside the pantry Pember has a 30-gallon keg of whiskey for her patients' use. Wilson's seen it there, but she's not about to give it up.

"You can't have the whiskey, Wilson," she shakes her head beneath the lantern, "and you *won't* have it; so you might as well leave."

"That's army whiskey, not yours."

"It's in my charge, and there it will remain. Now get out of my pantry; you are all drunk."

"Boys," an angry Wilson turns, "pick up that keg and carry it down the hill. I'll attend to her."

His companions, however, recognizing authority when they see it, no matter how small its package, aren't too sure. They shuffle uneasily at the door.

Encouraged, Pember stretches on her toes and attacks, "Wilson, you've skulked in this hospital more times than anyone I know. You know me; do you really think that you can steal anything I'm responsible for?"

Wilson tries again, "Now you just stop that talk! Your friends have all gone, missy; and we won't stand for your sass no more. Now get out of the way!"

Cursing, he steps forward and raises his hand to push her aside. He's barely moved when they all hear the sharp, unmistakeable click of a revolver's hammer being cocked. Then Wilson feels a very hard object jammed into his stomach.

"All right, Wilson. Leave! All of you leave! Right now!" Phoebe Yates Pember, all five feet of her, suggests. "And Wilson, if you think that a woman can't hit you with the first shot, remember that I have five more. It would be very hard for anyone to miss an ox your size six times. I *won't*, I promise you."

Stumbling back to the door, the bully tries once more, "You think you're brave, but you wait an hour or two. We may have pistols then, and you won't have it entirely your own way."

She raises the pistol. If he can't see her face he can't miss her weapon.

Again she orders, "Go, Wilson, go; and this time don't come back. The Yankees are here now, and I expect that they don't care much for deserters either. Go, before they come back."

When they've gone, she nails the top of an empty flour barrel across the door. Then, quite proud of herself, she sits beside her rescued keg of whiskey and, putting a candle, a box of matches, and her revolver within easy reach, falls asleep. Sleeps better than she has for days; awakened in the morning when one of her patients, concerned that she'd not appeared in the ward, comes to waken her.

A small victory, she smiles to herself, but we'll take whatever we can get.

GRANT

"Good morning, General," at dawn Grant is at Meade's headquarters when the ailing commander of his Army of the Potomac enters the room. "I hope your stomach's more settled now."

"Thank you, General. I think it is. I suspect that yesterday's excitement was a bit too much for me. At least my surgeon can offer no other explanation. And how are you?"

"Just fine," Grant as usual is very cheerful, "I've just had word from General Parke (Major General John G. Parke, commanding Meade's VI Corps). Lee's nearly cleared Petersburg, and Parke's men have entered the city."

"Then you were right to delay the attack we considered last night."

"It seems so. I figured that they'd evacuate the city during the night, hopefully Richmond too, and saw no need to sacrifice men for what we can walk into this morning."

"But there still is some resistance?"

"Scattered, sporadic. Mostly their sharpshooters trying to slow us down. A couple of them shot at me a few minutes ago, but your men drove them off. The last of their infantry units are backed up at the Appomattox bridge. I considered bringing in artillery, but I just don't have the heart to slaughter such tired, hungry, beaten men when there's a chance of capturing them alive."

He pauses, cigar smoke wreathing his face as he thinks. Then he quietly adds, "We'll pick them up in a day or so, General; I'm sure of it."

"Then you still don't want me to pursue them above the river?"

"No, not yet. Lee can't turn back. The only choice he has is to hurry west, just above the river, and try to get to the Danville Railroad ahead of Sheridan. Right now they're ahead of us, but by marching below the river we'll gain at least twenty miles on them. With luck, that'll be enough for Sheridan's cavalry to block their retreat somewhere near Burkeville. Stay in contact with their rear, and get some infantry up there to support Sheridan. Then have the rest of your army follow below the river as soon as things are settled here."

Their conference is interrupted by a staff officer with a captured Confederate officer. The Southerner speaks freely.

"I'm an engineer with the Army of Northern Virginia, General. Made up my mind that there's no point in more men getting killed in a useless fight, and I can help you end it. Lee's taken the army to new fortifications above the river. They're very strong; we've been working on them for months. If we cross the river here at Petersburg, I can take you to them."

"Wonderful," Meade exclaims. "A splendid opportunity, and we won't let them escape. Colonel Lyman, recall the units that already have begun marching below the river and send the others across here."

"Wait a minute, General, wait a minute," Grant gestures toward the map with his cigar. "Look at your map. Do you really believe that a general of Lee's stature would halt his army between the Appomattox and the James, knowing that we're fixing to get behind him?"

Smoke again clouds Grant's face as he studies the Confederate officer. Then, his eyes twinkling, he asks, "Are you a poker player, Major?"

"I've played the game, sir."

"Have you ever played 'Brag'?"

"No, sir. Understand it's an Old Army game."

"That's right," Grant, his mind made up, smiles again. "When this is all over and you see General Lee again, tell him that you did the best you could. But I think that you've been sent to throw us off the track."

Then he turns again to Meade, "The Danville Railroad's the only way Lee can go, General. We'll continue the march we've begun."

"But, if he's going that way, we can follow," Meade argues.

"Meade, you still don't understand," Grant frowns. "We don't want to *follow*. We want to get *ahead* of them. And this is the way to do it. Get ahead of them. Then, if Lee's still between the rivers, we'll have our net around him."

"Now," he ends the conference, "you have much to do. I'll get out of your way. Push your commanders; push them hard."

A bit later Grant has established his own headquarters in a large brick home on Petersburg's Market Street.

He's there, smoking his cigar on its porch, when a courier arrives with Sheridan's dispatch from Namozine Creek, some nine miles west of Petersburg. Sheridan's pressing hard against scattered resistance from Fitz Lee's and George Pickett's Confederate survivors.

At mid-morning Abraham Lincoln, accompanied by his son, Captain Robert Lincoln; Rear Admiral David Porter, Lincoln's escort; Lincoln's younger son, Tad; and White House bodyguard, William H. Crook, arrive.

LINCOLN

Lincoln also had risen early, eaten a hasty breakfast, then made his way to City Point's telegraph office where he updated his worn campaign map and, when Porter arrived, delighted in describing the situation.

"Now, Porter, here they are...and Sheridan's heading up this way....Well, you see that's gotta put Lee in a fix."

From time to time he'd paused to lift the three telegraph office kittens to his lap or to sit on the floor to laugh at their antics.

"You poor, little critters. What brought you to this camp of warriors? You say their mother's dead, Captain Beckwith?"

"Yes, sir," answered Grant's telegrapher.

The President's conversation revealed his thoughts, "Then she can't grieve for them as many a poor human mother grieves for sons who have fallen in this awful war, and who may yet fall if this surrender doesn't take place without more bloodshed."

He thoughtfully rubbed one kitten's soft fur and reminded it, "Ah, you little kitten. Thank God cats can't understand this terrible war."

"Get them some milk, Captain, and don't let them starve. There's too much starvation in this land."

Then a midshipman reported to the admiral.

"Sir, Vice President Johnson (Andrew Johnson) and Senator Preston King have arrived from Washington. They're on the *Malvern* and want to pay their respects to the President."

An angry Lincoln was firm, "Admiral, they'll only fuss with me for being here and get in the way of the work I'm trying to do. Tell them what you will, but don't allow them near me."

Porter nodded then instructed the young officer, "Tell Captain Barnes he's to entertain them. He might use the *Phlox* to show them the river, but he's to be very firm that the President is with General Grant and won't return until quite late. Meanwhile, we have no accommodations for them and suggest that they return to Washington with the afternoon's tide. Under no circumstances will they be allowed to get near the President."

Lincoln, as if the incident had not occurred, reached down to lift one of the kittens to where it could attack a pencil on the map table then quietly suggested, "Perhaps we should be on our way to Petersburg, Admiral?"

Porter would long remember the trip to Grant's headquarters.

As the special train of the Military Railroad was about to leave City Point, Lincoln's party was in its single coach, Porter on its platform. Then three well-dressed but rough-looking men called from the tracks.

We want to see the President!"

"No," Porter was firm, "you can't come aboard."

Ignoring his answer, two of the men tried to climb to the coach's platform. Porter reacted quickly, however, throwing both of them from the train. Before they could try again, the train pulled out.

The admiral brushed off his uniform, entered the coach, and quietly took his seat; relieved that the President apparently had not noticed the disturbance.

After a moment, however, Lincoln quietly looked up from his newspaper, studied the admiral over spectacles low on his nose, and smiled, "Admiral Porter, how much do you want for that trick?"

The admiral's problems continued at Meade's Station where Captain Robert Lincoln waited with a cavalry escort. The escort's commander ordered a sergeant to supply Porter with a horse, and Porter found himself with a raw-boned, hard-gaited, blind-in-one-eye, aged horse which, during the short ride, repeatedly tried to run away with him.

Several times the cavalry captain cast a baleful eye at his sergeant and suggested another horse for Porter; each time the red-faced but determined admiral rejected the idea.

As they reach Grant's headquarters an amused Lincoln asks, "Admiral Porter, you sailors know nothing of horses. Why in the name of heaven won't you rid yourself of that beast? His head's as big as a barrel. He's fourteen years old if he's a day. His hoofs cover most of an acre; he's spavined; and he's blind in one eye. Why do you stay with him? Get one of these soldier fellers here to pick you a good horse. Someone," he chuckles, "other than that sergeant."

"No, sir," Porter is firm. "No, I want *this* horse. And when we're done I mean to buy it and shoot it so that no one else will ever have to ride it again!"

They're all chuckling at Porter's determination when Grant rises from his chair on the porch, takes the ever present cigar from his mouth, and waves a greeting. Then he comes to meet the President.

Lincoln takes the brick walk with long, rapid strides, his face beaming with delight, to seize Grant's hand. He shakes it for some time as he pours out his thanks and congratulations.

"General, all along I had a hunch that you intended to end it right here; but I reckoned you'd have Sherman come up and help you."

"I considered that, Mr. President," Grant nods. "But Meade's army has fought Lee's army for nearly four long years. They deserve this victory; all by themselves. Besides, if the Western army came in now, there's a chance it could stir up some bad feelings between the two armies. We don't need that."

"I see," Lincoln replies, "and I appreciate your concern. I never thought of that part of it. I've been so anxious to get the job done that I didn't care where the help came from."

"Well, maybe I worried needlessly, Mr. President," Grant reflects. "Our soldiers get along well, and perhaps our politicans wouldn't have raised the issue either. I'm afraid I'm out of my element in these things."

"No," Lincoln chuckles. "I think that you handle politicians as well as you do your soldiers. Now, tell me what is happening."

They spend an hour together before Grant tells him that he must move his headquarters forward. Lincoln agrees.

"I'll just poke around Petersburg a bit then return to City Point. I'll be anxious for word from you because I really believe that we're about to end this terrible war. Can we do it without another battle?"

"I hope so, Mr. President. I hope so. When the time comes I'll make it as easy as I can."

As Grant turns Cincinnati west to catch up with his army, he glances back over his shoulder. Lincoln, from the rocker on the porch, is waving him on.

Later, as Cincinnati carefully picks his way between long columns of marching soldiers, Grant studies the land on either side of the slowly ascending road. It's a poor land. The sandy, deeply rutted road has been cut to pieces by the hooves of thousands of horses and the wheels of hundreds of cannon, caissons, and wagons. On either side of it are old, weather-faded, tumbled-down, split-rail fences or stone walls and tangled thickets of honeysuckle or Virginia creeper. Beyond them are piney woodlands; old, worn-out fields of tobacco, cotton, and corn; and sparse, overgrown farms with tired, weather-worn houses, for the most part abandoned.

"Poor country, Rawlins (Brigadier General John Rawlins, Grant's Chief of Staff)," Grant reflects. "Hardscrabble poor for certain."

"Yes, but there's the first spring flowers, Grant. See them? Violets, trillium, cowslips. Over there beyond the road. Julia'd like them. In a couple weeks the fields will green up, and that'll pretty things a bit."

"Humph," Grant puffs his cigar. "Julia'd like the flowers all right. For my part, I'd settle for a decent road or two. This one's too slow, too slow."

By mid-afternoon they've reached Sutherland Station, nine miles west of Petersburg. Burial parties still work there to clear the dead from the fight for the railroad. Not a pretty sight. And Grant, a lover of horses, also is troubled by the torn, swollen, wide-eyed bodies of horses dragged into piles for burning.

At Sutherland Station he receives another report from Weitzel:

We took Richmond at 8:15 this morning. I captured many guns. The enemy left in great haste. The city is on fire in two places. Am making every effort to put it out.

Passing it to Rawlins, he orders, "Read it to the staff, General; and be sure that the news gets to the men. I'm sorry we didn't get it before we left the President, but I suppose he's heard it by now."

_____

At dusk he halts his staff.

"I'd like to push on, gentlemen, but we're losing our light pretty fast, and the road's too packed with troops to go far in the dark. The rations didn't get up today, Colonel. See what you can do about it."

"No one seems to mind, General. The men are all for pushing on."

"I know. All they needed was one great victory. Now they've had it, and they see the end in sight if we can get ahead of Lee's army. But tomorrow they'll be hungry; get our 'cracker line' open."

_____

There's word from Sheridan, an encouraging report of a routed army:

The enemy threw their ammunition...(beside) the road and into the woods...then set (it) afire....The woods are strewn with burning and broken down caissons, ambulances, wagons and debris....I have taken about 1200 prisoners...and all accounts report the woods filled with deserters and stragglers.

_____

During the evening Grant quietly studies his maps, ponders the next day. He'd ridden beside Meade during the afternoon, but Meade, still suffering from nervous indigestion, has retired early.

"There's an opportunity here," Grant puffs his cigar and mutters to himself as he traces lines on the map with a hand that has known heavy work.

"An opportunity, but it won't last; and it all depends on the next two days. If Lee's to hit the Danville Railroad, Burkeville's his best bet. Sheridan ought to be able to get there ahead of him, but he'll need infantry to stop him."

"He has Griffin's V Corps," he counsels with himself. "I expect," he smiles, "that Sheridan and Griffin are getting used to each other by now, and Griffin'll push hard to keep up with the cavalry. Griffin's a good man; probably'll have Chamberlain (Brigadier General Joshua L. Chamberlain) setting the pace. Must keep after Meade though. He can't seem to understand what I'm trying to do. In the morning I'll have Meade stay near Griffin. If there's an infantry fight, it'll likely be there. I'll ride with Ord. Maybe, off on his own, Meade'll do better. We'll push hard for Burkeville."

He sends word for the troops to march at 3 a.m. Then, still thinking about the tactical situation and trying to visualize the roads, the problems, the fighting yet to come somewhere up ahead, he too drops off to sleep.

LINCOLN

In Petersburg, after Grant rode off to catch up with his armies, Lincoln and his party ride through the city. Except for throngs of Negroes crowding the streets, shouting, singing, grinning at their Yankee liberators, Petersburg seems deserted.

Porter and Crook, Lincoln's personal bodyguard, nervously scan the shuttered windows bordering the cobblestoned streets. Any one of them could hold a Confederate sharpshooter left behind or an assassin.

Once Crook, detecting a fluttering curtain, draws his revolver and calls out, "Mister Lincoln, Admiral, Careful there."

Then the curtain parts more, revealing a small, pale, wide-eyed boy.

Abraham Lincoln responds first, "Wait, Mr. Crook. Wait. Why it's only a child. No older than Tad, and he means no harm."

Then a woman, frightened, appears at the window, snatches the boy away.

"Can't be too careful, Mr. President," Crook reluctantly holsters the heavy revolver.

"I know," Lincoln smiles sadly, "and I appreciate your concern. These are terrible times; but I can't believe that anyone would intentionally harm me."

Several times hordes of Negroes close around the presidential party, and Lincoln must say a few words to them before passing on.

One of them, a white-haired old man, throws himself at Lincoln's feet.

The President, embarrassed, reaches down to help him up, shakes his hand, tells him, "You are free now; you must never again bow to any man."

Petersburg's tobacco warehouses have been thrown open and soldiers are helping themselves to long leaves of tobacco. Porter ties several bundles of them to his saddle; and the President and Tad Lincoln take several for souvenirs. There is no widespread looting, however, and Lincoln is pleased that his Union soldiers are behaving well.

Regiment after regiment of Parke's corps hurry past to catch up with the army west of Petersburg. Recognizing Abraham Lincoln, soldiers cheer him.

A regimental sergeant major waves his hat and shouts, "Three cheers for Uncle Abe, boys!" and the cheers echo in the narrow streets.

Another soldier waves and, as if they were old acquaintances, calls out, "We'll get 'em, Abe! You go home and sleep sound tonight. We mean to put you through!"

On the train returning to City Point Tad spots a large turtle sunning himself beside the track. Lincoln has the train stop while Tad, in his cut-down soldier's uniform and encouraged by the cheers of their soldier escorts, chases down his quarry.

At City Point Grant's telegrapher is waiting with another scolding telegram from Edwin Stanton, Lincoln's anxious Secretary of War, in Washington:

...Allow me respectfully to ask you to consider whether you ought...to expose the nation to...any disaster to yourself in the pursuit of a treacherous and dangerous enemy like the rebel army....Commanding generals are in the line of duty running such risks, but is the political head of a nation in the same condition?

Lincoln smiles and scribbles a reply:

Thanks for your caution, but I have already been in Petersburg. Staid (*sic*) with General Grant an hour and a half and returned here....Richmond is in our hands, and I think I will go there tomorrow. I will take care of myself.

"Captain Beckwith," he smiles, "please send this off 'lest Secretary Stanton's milk sours for sure."

As he returns to the *Malvern*, he pauses to watch a Pennsylvania regiment herding Confederate soldiers to the prisoners' compound. A large, heavily bearded Union major is shouting a final offer.

"Boys, this war is over! All who'll take an oath of allegiance to the Union, right now, step out here. It's that or the Bull-Ring behind you. Come on now, take the oath and we'll ship you to Washington for detainment. Not a prison, just detainment. This very afternoon; with full protection. Now, who'll take the oath?"

As Lincoln watches, not one prisoner steps forward. Not one. The major, with less enthusiasm now, as if he's been through this before, repeats the offer several times.

Finally, here and there from the ranks of prisoners, about a hundred ragged, lean Confederate soldiers, heads slumped as if they are terribly ashamed, hesitantly step forward.

The other Gray prisoners immediately turn on them with boos and catcalls: "Traitors!" "Cowards!" "Galvanized Yankees!" until Northern guards hurry the small group toward another, equally small, group of Confederates waiting to board a small steamer at the dock.

"About a hundred, Porter. No more," Lincoln shakes his head. "Must be 5,000 of them and, after all they've been through and knowing that the end must be in sight, only five score willing to surrender! That shows us what Grant's facing out there. They'll fight as long as they're able to fight. And more blood must be shed. Too bad, too bad."

LEE

Above the Appomattox, Robert E. Lee also has spent much of the day assessing his army.

They begin their march in good spirits, Lee astride Traveller, waving to his passing regiments: alert, cheerful, animated, commanding. Once he shares with his aide, Colonel Walter Taylor, just returned from a last-minute trip into Richmond to be married, "I have my army out of the trenches now, Colonel. To follow me Grant must leave his lines, and he'll get no more benefit from the railroads or the James River."

Lee's headquarters' staff, also happy to be on the move again, tease Taylor about his brief honeymoon. Tease him but more discreetly when Lee is near. Some still recall a railroad coach two years before when one of them began to tell a somewhat off-color story then looked up to see that Lee had returned to the coach and is listening. Lee's normally ruddy features were more flushed than usual, and the teller, realizing that he's on uncertain ground, nervously ventured, "It's all right, General. There are no ladies present."

Lee, however, did not smile, only quietly rejoined, "No, sir, but there are gentlemen present."

They never did hear the end of that particular story, and even now are prudent around him. West Point classmates had not called Lee "the Marble Statue" without reason.

This morning, however, he's in good spirits but reminds them.

"Provisions. They're critical if we're to continue. Our men are hungry, but they can hold out another day by sharing what they have, just as they've always done. But we *must* reach Amelia Court House tomorrow. Our rations and ammunition will be there."

In the early afternoon Lee's staff happen upon an unexpected meal.

A dozen miles from Petersburg is Clover Hill, the country home of Judge James M. Cox. Clover Hill: a rambling house, studded with gables and dormers and wrap-around porches softened in the summer by hundreds of roses, is known for its hospitality.

Learning that Lee's army is passing nearby, Judge Cox sends a servant to invite Lee, Longstreet, and their officers to join him at Clover Hill for lunch.

Arriving, they find the house bustling with excitement as Cox's attractive, gracious daughter, Kate, has the meal prepared.

The young girl falls under the spell of Lee's quiet dignity but is concerned about him.

"General," she tries to reassure him, "we'll yet win. You'll join General Johnston's army in North Carolina, and together you'll win."

His smile is warm, grateful, but weary. "Thank you. I hope so. But, whatever happens," he tells her, "no men ever fought better than mine."

Served a mint julep, he asks instead for a glass of ice water. Later, during the meal, he eats only enough to be polite and, when asked, "General, do you prefer cream with your coffee?" he confesses, "Yes. Cream or milk. The truth is, I've not had true coffee for so long that I wouldn't dare take it in its original strength."

Kate Cox, busy cutting Longstreet's food (he still recovering from the bullet wound he'd taken in the Wilderness), asks, "General Longstreet, is the food all right?"

"Oh, yes, ma'am. Excellent."

"But General Lee has hardly eaten a bite, and now just a sip of coffee. Is he not well?"

"He's tired, ma'am, but he's all right. He'll eat better tomorrow when he knows his army's been fed. As for his coffee, he sends all his coffee to our wounded in the military hospital." Then he adds, "'Preciate your concern, ma'am, but it's best to let him be."

Lee interrupts with a broad smile. "Thank you for your hospitality, Miss Cox, but we must get back to the men."

She watches as Lee and his staff ride down the long, tree-lined lane. She'll never see that lane again without recalling that magnificent horseman, his red-lined gray cape thrown back over his shoulders; erect, commanding, confident, returning to his army.

It's good that Lee's had that brief rest because by mid-afternoon the situation his army faces has deteriorated.

Couriers report that a pontoon bridge, supposed to be laid across the river, is not there. They'll use planks from a barn to build another bridge, but it will take time. And high water has washed away a second bridge. Troops must march farther or rebuild it too. Hard work, particularly when a man is bone-tired.

Straggling has increased. Following rest stops exhausted riflemen and walking wounded are slow to take up the march again, often losing their regiments entirely. Officers and sergeants herd them into the long column, anywhere, until the army can be sorted out in the evening.

Lieutenant J. F. J. Caldwell's South Carolinians, men who've always kept up on the hardest marches, have begun to straggle, moving without order, at no regular pace. When a man becomes too tired to walk, he simply drops beside the road until he can pull himself up by his rifle again. Then he shuffles forward toward where he last saw his regiment. Sooner or later, he reckons, he'll catch up.

Luckier soldiers are able to scrounge scraps of food along the way. Those who have usually shared with those who don't, breaking a rock-hard, worm-ridden hardtack biscuit or a small slice of rancid bacon in half or sharing closely watched sips from a canteen of cider or molasses.

Privates Timberlake and Warren of the 2nd Virginia Infantry Regiment count themselves blessed. Between them they have a pound of coarse cornmeal and several small strips of bacon.

Told they'll be stopped for awhile, they build a fire and make a batter of water and cornmeal. Then they pour the batter into their frying pan, the remains of Timberlake's tin canteen.

Their meal is just beginning to take shape when a bugle summons the 2nd Virginia back to the road.

"Oh, God, Marshall," moans Timberlake. "We gotta git. I'm gonna have to throw this away!"

"No, don't do it," his messmate is firm. "We'll eat the damned thing anyway, just as she is."

An unusual blessing before their meal, but it must do. Each swallows half the batter, and Timberlake licks the "skillet" clean. Then they run to catch up with their rifle company.

They'll be grateful that they've had at least that much; for the next two days all they'll have to share are several handfuls of sassafras buds and one ear of parched corn.

Late in the afternoon, Lee's soldiers, known for singing and cheering on the longest, hardest march, have little to say, let alone sing. Instead they shift their rifles and blanket rolls to ease cramped muscles, lean into the road, and focus their half-closed eyes and hazy thoughts on the men ahead of them. So long as they move, those behind try to keep up. When a man stops for more than a moment or two those behind him ask no questions, just sink to the ground themselves; grateful for some rest.

Lee's army, from private to general, is weakening. The long months in cold trenches; the little food and that poor at best; and the hard fighting of the past few days have all taken their toll. They'll make about twenty miles this day and lucky to do that.

Exhausted or not, they're mindful that their enemy are marching too. Scouts report that the sandy road that leads to Amelia Court House, however, is clear. And, across the Appomattox, the few Federal cavalry patrols that Lee's flank guards see are too far away to do harm. For the moment, they sense that the greatest danger to their army is from their rear, and there Gordon reports that their pursuers are hanging back. The Yankee army seems to have mysteriously disappeared and, when some enemy soldiers appear from time to time, they seem uninterested in taking up the fight. Still, they're out there, and Lee's soldiers know that.

Lee, establishing his camp for the short night's rest, is grateful that they've not yet threatened his march. And he's heard from more of his scattered units. With luck, he'll have all his army, what's left of it, together in the morning. Then a hard march and, by mid-afternoon, Amelia Court House and the desperately needed rations.

Colonel Marshall, coming to Lee's tent to deliver a message, finds his general leaning forward on his camp chair, his eyes closed, his lips silently moving.

Marshall quietly lets the tent flap fall. His message can wait.

All day Major Edward M. Boykin's 7th South Carolina Cavalry Regiment guards the rear of Lee's still-divided army.

The road ahead of them is strewn with the debris of retreating soldiers: blanket rolls, canteens, worn-out shoes, pots and pans, small cook stoves. Cannon, axle deep in mud, spiked and abandoned beside the road; broken caissons and burning limbers. Some artillerymen are busy placing ammunition box headstones inscribed with fictitious names of soldiers and units over "graves" that hold cannon that must be abandoned. Perhaps they can be retrieved later.

Boykin's men make eleven miles this day, and lucky to do that. At dusk they camp near the road where he divvies up a little parched corn for the horses and last bits of bacon and hardtack for the men.

"Amelia Court House," they reassure each other. "Ought to be there by late morning. Plenty of food then."

STILES

For Major Robert Stiles, the pious commander of a Confederate artillery battalion who constantly worries about his men, the day's march is very long. His gunners, accustomed to quiet garrison life on the James River, aren't hardened to hiking long infantry-miles. Many are teenagers, or gray-haired veterans, or soldiers so badly wounded that the doctors deemed them fit only for quiet duty. That is, until yesterday. Now they're marching beside young infantrymen, trying hard to keep up. Pride helps them; pride and heeding what Stiles and their sergeants have been telling them about thinning the loads they must carry on their backs. After every rest stop their blanket rolls lose more of their former contents; with "necessities," no longer important, left beside the road.

Stiles will always remember their last view, from a hill above the Appomattox, of Richmond burning behind them: explosions and heavy, dark clouds, periodically shot with flames, hovering over the fallen city.

Then he quietly orders, "Move your batteries forward, gentlemen; move them forward."

All day he walks in the dusty rear of his battalion, prodding stragglers and using two horses to relay exhausted men ahead then return for another load. In this way his footsore gunners are managing to keep fairly closed up.

––––––––––––

During a rest halt Stiles walks to a ramshackle farmhouse porch to talk with an old man whose wife and daughter, a soldier's widow dressed in black, are watching Lee's long column pass.

"Your sleeve's torn, Major," the young woman touches his arm, "let me mend it."

As she works, a second girl, pretty but red-eyed, weary, disheveled, also comes to the porch.

"Mother, if Bob comes with these men and doesn't mean to stay right here, he's no husband of mine. I don't want to see him. Not now, not ever. Tell him that."

Stiles, realizing that his soldiers are listening intently, challenges her.

"Ma'am, if you tell him that, you're asking him to desert. Surely you don't mean to do that."

"Oh? He's my husband, Major; at least until now. I'll do as I wish."

"Yes, ma'am, that's your right for certain. But my soldiers and I belong to the same army as your husband and, the way things are right now, none of us needs to hear that kind of talk."

"Army! You call this mob of cowards an army?" the distraught woman lashes out. "All day I've watched you pass. Tails between your legs! Leaving us to face the savage wolves who are right behind you. Cowards! My husband's place is here protecting his family."

"Ma'am," Stiles' voice is reassuring, pleading for understanding, "these are my men, and I know them. They're not cowards; they're good soldiers. We're not about to surrender to the Yankees. Somewhere up the line, General Lee'll turn us around, and we'll fight again. Not here but somewhere."

"Talk's cheap," she laughs bitterly. "The Yankees maybe over that next hill, and those of us who must stay and face them have no army and no country to protect us. All gone. My husband's worn that uniform of yours more'n three years. Given it all he had. Now his place is with me and his starving children, not following some lost cause. With us. And if he doesn't come now when we need him, he needn't come later."

The weary, bedraggled, but dedicated major tries another tack as his watching soldiers listen.

"What regiment's your husband with, ma'am?"

"The Stonewall Brigade, sir!" she straightens, her eyes flash and her voice, for the first time, is strong, firm. "General Stonewall Jackson's old brigade. Or what's left of it!"

There is deep pride in her voice as she adds, "Since '61. And he's seen the worst of it."

From her apron she draws a wrinkled, torn, faded, many-times studied piece of paper and hands it to him. Proof of her claim. Carefully unfolding it, Stiles reads the tear-stained words on a worn furlough pass signed by Robert E. Lee himself, and awarded for her husband's courage a long time ago.

"Extraordinary gallantry in battle," Stiles reads, and his mind pictures the desperate action the words represent.

Turning to his soldiers, he raises his voice just a bit, "Listen to this, boys." Then slowly, deliberately, he reads the furlough pass, emphasizing Lee's personal signature at its end.

When he's finished, the young woman, reaching to retrieve the paper, chokes on sobs she no longer can control.

"This little paper," he folds it carefully, "is your most precious treasure, isn't it?"

"Yes, sir; reckon it is. It's all I have."

"Ma'am, my men and I understand that more than anyone else ever could. But don't you see that if you ask your husband to leave his regiment, to stay here with you, you'll turn him into a deserter? And you know that he couldn't stay on for more than a few hours, even if he wanted to. Our rear guard's picking up stragglers and deserters. They'd take him from you; not a brave soldier but a prisoner. Then this priceless paper would be only a bitter reminder you'd all have to live with for the rest of your lives. You're going through a hard time, but you musn't let that happen. Not after all this."

He stops, his argument exhausted. For a moment the only sound is the woman's labored breathing. Then she carefully returns the paper to her pocket. Another pause then, wiping her eyes with the hem of her apron, she turns to her mother.

"Mother, I've changed my mind. If Bob passes, tell him to go on; keep going. Finish the job he set out to do. Then he's to come to us soon's he can."

As she turns away Stiles' men cheer her, and all of them feel a little stronger for the march.

At dusk they share their last rations in a clover field just off the road.

CLAIBORNE

Doctor John Claiborne's little Confederate hospital train left Petersburg with Longstreet's column and is miles down the road when Claiborne realizes that his Scottish terrier, "Jack," is missing.

"That dog," Claiborne explains to his surgeons, "isn't much as dogs go. But he likes me, and he'd follow me anywhere."

Then he reflects, "Well, almost anywhere. It's just," he continues, "that he's got too much pride to walk. That dog's ridden farther, fallen out of more carriages, and been run over more times than any other dog on earth. Now he's gotten himself lost. He's probably looking for us right now, and we must find him."

"Romulus," he calls to the young slave committed to his care just before the retreat began, "you hustle back to Petersburg, find Jack, and carry him here. We'll be somewhere up this road. You find us."

Romulus, not at all sure that that's what his mother had in mind when she told him to follow the doctor to the ends of the earth, starts to protest then gives it up, turns back down the long road they've just traveled.

Claiborne's surgeons bet that they'll never see boy or dog again. That evening, however, they're sitting before their campfire when a grinning Romulus walks into their camp. Trailing behind him, on a leash made of knotted handkerchiefs taken from some abandoned Petersburg store, is Jack. Romulus had found the dog sitting in front of the store, scooped him up, then fled as "Yankee soldiers, massah, a whole passle of 'em, come down de street."

Jack seems tolerably glad to see the doctor, sharing a hardtack biscuit then crawling under Claiborne's blanket.

About midnight, however, a tremendous explosion shakes their little camp. Ammunition being destroyed, without warning, in a nearby field. Jack, terrified, leaps from the blanket and, without even a backward glance, disappears into the night, loping, as fast as his dignity will permit, back down the Petersburg road.

LIEUTENANT JOHN WISE

About eighty miles down the Richmond-Danville Railroad from Lee's army, Lieutenant John Wise, 18-year-old son of the former Virginia governor, Brigadier General Henry A. Wise, an officer in one of Fitz Lee's cavalry detachments, is guarding tiny Clover Station.

They've been there for weeks; a boring assignment, particularly for a young officer who wants to see combat but knows that Clover Station's too far beyond the Richmond-Petersburg lines for even an adventurous Yankee patrol to reach them. All that, however, changed two days ago with the news that Grant has broken Lee's lines below Petersburg. Like a dark storm cloud, moving closer and announcing its coming with wind and big drops of rain, fleeing refugees and the constant clicking of the telegrapher's key alert them that action is coming their way. Coming their way fast.

Wise and his handful of cavalry troopers haven't missed the signs. So he's posted two men on each of the roads leading to Clover Station and along the railroad itself. Meanwhile, he remains at the station, the chattering telegrapher's key his link with the world.

Last night a half dozen trains came through. Each it seemed in worse shape than the one before it, but somehow puffing and clanking and limping toward Danville. They pause at Clover Station for wood and water, and during their wait government officials come to their rear platforms to speak to the crowd that quickly gathers below them.

"Yes, it's true. Pickett lost badly at Five Forks, and fighting's very heavy below Petersburg. But," they reassure their listeners, "the army's not beaten. Not so long as we have Robert E. Lee. If he must pull back, he'll do it in good order. Then, when he's drawn the Yankees out of their lines, he'll turn and whip them. Just as he's always done. You just wait and see!"

Wise, hearing this several times, feels better until he begins to note the contents of the trains. One carries the Confederate treasury. Others have the records and the clerks of the various departments. A fleeing government on wheels.

Nearly half a day after it left Richmond (normally a four-hour run) the presidential train: dirty, rusting, leaking, its paint peeling away in long strips, arrives. It has broken down several times and wasted several more hours when Yankee cavalry were reported cutting the tracks ahead of them. If that were not enough, the smoke and cinder-filled, back-breaking coaches have caused their passengers to feel every mile of the miserable trip.

Lieutenant Wise sees his president, Jefferson Davis, waving a hand to the small crowd but Davis, seeming physically and emotionally exhausted, remains seated, not coming to the platform to speak to the gathered crowd.

Wise enters the coach to find that the only good cheer there hovers over a group of men huddled about the wife of the treasury secretary, George Trenholm. They seem to be enjoying a picnic of some sorts which she is drawing from several large wooden hampers.

Wise can think of little to say to anyone on the train and, when it puffs out of the station, he stands alone on the tracks, relieved to see it limp out of sight.

There are several more special trains. More clerks, more archives, more signs of a government fleeing the Federal army. Each successive train, it seems to Wise, more disorganized and bizarre appearing than the ones before. As if frantic Richmond trainmasters were desperately trying to unload the last of their responsibilities. In one car: many empty seats but a cage with a parrot and a box of tame squirrels, escorted by a hunchbacked man.

Then, from the last train puffing out of the station, a man calls to the young officer, "Richmond's burning, Lieutenant. Gone. All gone. Goodbye. Goodbye."

Wise, standing alone on the tracks, watches the train vanish from sight. Then he reaches down to touch a rail, the iron ribbon his last contact with the train and, he feels, with a dream. When the rail stops vibrating, the contact is broken. For a moment Wise almost regrets not having jumped aboard that last coach. Then, ashamed of his thoughts, he turns back.

"Sergeant Brown, that's the last one. Tell our boys to look sharp; we don't know how far the Yanks are behind them."

DAVIS

When the presidential train reaches Danville, a waiting crowd greets Davis, and he establishes a temporary White House in a Main Street home. Then he sets out to inspect Danville's defenses, trenches that he anticipates Lee's army will occupy when they arrive.

He's not happy with what he finds. Trenches and earthworks, he decides, are in the wrong places and are nowhere near done. Food and ammunition he'd ordered have not arrived. Worse yet, there's been no word from Lee. Davis has no idea what is happening to the general commanding all his armies.

Rear Admiral Raphael Semmes, famed former commander of the Confederate raider *Alabama*, reports to him there. Having scuttled the James River fleet at Richmond, and having no transportation for his beached sailors, Semmes commandeered an abandoned locomotive and cars from a Richmond siding. Then his sailors fired its boilers with wood from fences and buildings and loaded the improvised train with all the refugees it could carry. Now he's at Danville, awaiting orders.

"Admiral," Davis tells him, "you are now a brigadier general. Take charge of the city's defense lines. Come, let's see them together."

So Brigadier General-Rear Admiral Raphael Semmes takes on his final assignment for the Confederacy.

SHERIDAN

Below the Appomattox, Union General Philip Sheridan pushes his cavalry and General Charles Griffin's V Corps forward against scattered but stubborn resistance from the remnants of Pickett's and Fitz Lee's commands.

It's a poor country with crude roads and hardscrabble farms. Roads and crossroads, that suddenly are important, must be taken to assure the advance of the army and the vast trains of supply wagons behind them. Where the passage has been argued, exploding shells and Minie balls have torn spring-new leaves from trees. Wounded men sit beside the road or filter back toward the hospital trains. Bodies are hastily buried in shallow graves that everyone knows the rain will wash out or animals will find. The dead are buried pretty much where they fall while the rest of the column moves on, pretending not to notice.

Cavalry will outmarch infantry for a day or two, but by the third day the horses must be rested. Not so with the foot soldier. So Griffin's infantry is keeping fairly close to Sheridan's cavalry and, for now at least, the little general is satisfied.

Near Namozine Creek Custer's troopers come across a crossroads which Fitz Lee won't give up without a fight, a running, hard-slashing fight. Captain Tom Custer, younger brother of Sheridan's fiercest fighter, Major General George Armstrong Custer, loses three officers and eleven men in the fight but personally captures a Confederate battle flag and for it will be awarded the Congressional Medal of Honor.

Throughout the day Union Major Henry H. Young's scouts also have served Sheridan well.

Young, in his mid-20s, slender, muscular, rather shy among strangers but a reckless daredevil on a cavalry raid, had been noticed by Sheridan who then ordered him to volunteer to command a similarly "volunteered" group of cavalrymen. They'll be Sheridan's personal scouts, operating much of the time behind Confederate lines, doing whatever they can to disrupt Lee's army.

Young then selected about sixty Union troopers for the task, equipped them with Confederate uniforms and equipment, and taught them to speak with Southern drawls and casual familiarity about the enemy units and commanders they're likely to encounter.

He warned them that, if they're caught, they're likely to be hanged, but they shrugged their shoulders at that news. When you're in your early 20s, and there's a chance for adventure of that sort, some men don't mind the fine print.

For months they'd been very successful, ranging far ahead of Sheridan's regular cavalry. Sheridan's scouts: sometimes dressed as Confederate officers or sergeants; other times wearing faded, patched country clothing and riding decrepit horses or mules with makeshift bridles and saddles: displaced farmers or roving blacksmiths. Either way they ride through Southern picket lines and cavalry cordons; deliver bogus messages to Confederate headquarters; pause to chat around their enemies' campfires; spread confusion liberally.

Most of them get back alive, and their information allows Sheridan's cavalry to slash at isolated Confederate units: drawn-out artillery batteries or wagon trains. The scouts who are lost, well, good men but gone now. So who's ready to take their places? Young always finds new volunteers.

There are compensations of course. Riding ahead of the rest of the Union army, Young's scouts explore the whole network of country roads, lanes, and cowpaths. That means they reach Southern henhouses and barns far ahead of other Yankee soldiers. And they aren't above helping themselves to cash, or jewelry, or other valuables from homes they find; most often leaving their victims with the impression that they've been robbed by Confederates, not by Yankees. A wild crew. Every army in every war has found units like Young's scouts of great value. And they fear no one or no thing except their soft-spoken leader, Major Henry Young, or the black scowl of Phil Sheridan.

This evening, falling back after a running fight with Custer's men near Namozine Church, Confederate Brigadier General Rufus Barringer and about 300 of his North Carolina cavalrymen are met on the road by several Confederate guides who identify themselves as from Major General Rooney Lee.

"Evenin', General Barringer," their sergeant drawls. "We've been sent to lead you into camp for the night. Good ground and water nearby. You're to wait here 'til mornin' when the General will have orders for you. If you'll just follow us."

Exhausted from the day's fight, Barringer and his men follow. The campgrounds *are* excellent. Barringer and his men are grateful; they don't often have personal guides, especially good ones like this.

"No need to post guards, sir," the sergeant volunteers. "You've had a mighty hard day. My men and I are pretty fresh. We'll keep an eye out for you; and in the mornin' you'll have your orders."

The Tarheel troopers rest quietly, their hosts watching just beyond their campfires. Then their rest is rudely interrupted. Union cavalry have surrounded their camp, and their soft-spoken Confederate sergeant, now with a definite Yankee accent, rouses Rufus Barringer.

"General? Reckon you'd better get up, sir. Got our Yankee cavalry all around you. And those orders I told you about? Well, they're from General Sheridan and Major Young. Seems we work for them, and they figure it's best if you just come along quietly now, without a fuss."

CHAPTER 2

The Missing Rations   April 4, 1865

THE MARCH

In the early morning twilight Meade's Army of the Potomac and Ord's Army of the James march west below the Appomattox. Ahead of them Sheridan's cavalry and Griffin's V Corps' infantry are skirmishing with retreating Confederate survivors of the battles for Five Forks and the Southside Railroad.

Above the river Lee's army, still without Lieutenant General Richard Ewell's Richmond command, also is marching west, toward Amelia Court House where Lee hopes to unite his army and to feed its men and animals.

The roads the armies travel, seldom more than narrow dirt tracks, wander through scrub pine forests, fallen timber, and thorny blackberry bushes; cross one creek after another; follow red clay ridges or cut across mostly abandoned fields of wheat, tobacco, or corn.

The roads have been flooded by spring rains. Where the mud and water are axle deep on wagons and cannon, and the horses and mules, even when double teamed, can't pull them free, infantrymen add their strength to the wheels. Sometimes it works; sometimes it doesn't. When it doesn't, the horses and mules are led aside while soldiers corduroy the road with logs. Then the heavy loads can move forward. All of it takes time.

The Appomattox, still separating the armies, curves often, its waters dark, cold, and deep. Along its banks young trees struggle to reach the spring sunlight while old ones, sighing with the spring winds, bend from the weight of their years and the tangled masses of wild grape and Virginia creeper coiled around their branches: strangling dark-brown vines that double back on themselves until whole trees seem tied to the earth with heavy ropes.

There are signs of spring: buds not yet fully coaxed out by the spring sun, wildflowers, patches of green covering the winter-brown earth. Mostly, however, there's still a brooding winter grayness to everything.

The marching soldiers don't have much to say this morning. It's cold and wet, and the road steadily rises; best save their strength. From time to time drummers beat a half-hearted cadence, but for the most part they too simply slog along, trying to think of anything other than the miles ahead of them.

Now and then a soldier falls out, waving, "You fellas keep goin'. I'll catch up when I can." Then he sits beside the road.

At the rear of the army provost guards pick up stragglers and would-be deserters. Provost guards, who must march in the dust of the army all day, aren't a particularly happy group. So they don't deal kindly with suspected malingerers or deserters. If a soldier drops out, it's best if his feet are bloodied.

A flock of low-flying ringnecks is startled by a long column of soldiers emerging from a heavily wooded stretch of road. Their leader honks an alarm and swerves up and away. The rest of the flock, knowing about the long sticks the soldiers carry, take up his cry.

Now and then the raucous cry of a crow, or the rhythmic sound of a distant cowbell, or of an axe chopping wood also breaks the silence. The sound of chopping is sure to alert the marching soldiers. It could be a farmer clearing his land, but it also could be trees being felled across the road ahead or earthworks being prepared; either way the first hint of a fight.

From time to time narrow, tree-bordered, rutted lanes lead off from the road, tempting stragglers. Somewhere up those rutted lanes there may be a house or two. But this morning, unless a man's awfully hungry or thinking abut deserting, isn't the time for exploring byways. Sergeants are even grumpier than usual: herding their squads along, sometimes by good natured cajoling but more often by promising to spread-eagle stragglers over wagon wheels when the march ends, to dunk them in the nearest creek or, if their standard threats seem not quite enough, the always ominous, "Just you wait and see!"

Confederate sergeants have a simple message to motivate their men: "There's food up ahead at Amelia Court House. The quicker we get there, the quicker we'll eat." And they needn't remind anyone, "You sit there a little bit longer, boy, and you'll find yourself back at City Point. You can bet the Bluebellies'll be comin' up behind us any time now."

Confederate sergeants fear that if a man wanders off he's gone for good; Yankee sergeants, on the other hand, aren't worried about desertion, not in the heart of enemy territory; but they don't want their men looting or getting themselves killed by some bushwacker.

Union veterans warn younger soldiers, "Don't go wanderin' off 'lest you'd like your throat cut. Stay close to the rest of us. 'Sides, nothin' up that lane's worth lookin' at. Been with this army since '61, and I reckon I've been up a thousand of 'em; ain't found one worth walkin' yet. You stay close at hand. Soon's we can get ahead of Johnny we can end this thing."

A simple strategy; both armies have grasped it, and sergeants have less prodding to do on this march than any other they can remember.

Infantrymen, Blue or Gray, lean into the road, fix their eyes on the backs of the necks of the men ahead, and match their steps to their sergeant's occasional rhythmic grunting, "Keep movin'; left, right, left, right; keep movin'." A hard pace.

The cold, overcast, drizzling morning causes soldiers to think about how nice it'd be to have a campfire, and hot coffee, and maybe some "skillygallee," or "Johnny Cakes", or "cush." "Cush": bacon fried in a half-canteen skillet then stewed with cold beef, water, cornmeal or biscuit or hardtack, and all of it stirred with a stick or bayonet. "Cush": a real delicacy when a soldier's hungry. Yes: hot coffee, cush, and an hour's rest. Something to think about.

"Stick it out 'til sunset," Yankee soldiers tell each other; "then the wagons'll be up."

"Amelia Court House," say Lee's veterans. "Amelia Court House. Maybe just over that hill up yonder. Keep movin'!"

Grant is with Ord's leading regiments. Close enough to occasionally draw sharpshooters' fire. When that happens, he studies it a bit, shifts his rain-soaked, dead cigar in his mouth, then nudges "Jeff Davis" forward again. As usual, he doesn't have much to say, just keeps a slow, steady pace west. Closer each hour to the critical railroad.

MEADE

General George Meade, still ill, is forced to ride an army ambulance again today. A bouncing, jolting, agonizing ride northwest toward Jetersville where messengers say Sheridan's trying to bring on a big fight. The reports from Sheridan this morning are all optimistic. And also up ahead, Griffin reports that, Sheridan's horses having slowed a bit, V Corps' infantry are catching up now. Closing to supporting distance if Sheridan gets himself into trouble.

Grant had ridden for a time with Meade before he swung away to find Ord's army. He's not been gone long when Meade's headquarters' band unexpectedly swings into "Hail the Chief." Meade, startled at Grant's surprise return, appears at the canvas opening of his ambulance.

Sure enough, Grant and two or three other men are nearing their crossroads. Grant smiling and acknowledging the bandmaster's salute with a grin and a wave of his cigar. Strange, Meade thinks, first time I've seen Grant pay a lick of attention to such tomfoolery.

Then the three riders draw nearer, and Meade realizes that "Grant" is Colonel Mike Walsh, an irrepressible Irishman, once a sergeant, often mistaken for General U. S. Grant. Walsh once more taking advantage of an eager bandmaster's premature salute.

Meade, his upper body clear of the ambulance's cover, starts to shake his fist at "General Grant," then realizes that he and "Colonel Mike" have been down this road too often in the past for Meade to change him now.

The irascible general's fist becomes more a wave, the scowl the closest Meade can come to a grin. "Colonel Mike" sees it and acknowledges Meade's surrender with a salute of his own: a big cigar between the fingers of his right hand, the hand salute more a flip wave than regulations would permit, and the Irishman's cheerful, "Mornin', General darlin'; mornin', and a good day to you, sahr!" Then he's ridden by, heading somewhere up the way.

Meade, chuckling to himself, suddenly realizes that he has no idea where the man has been or where he's going but, he shrugs his shoulders, "I suppose it's important."

––––––––––

At sunset Meade establishes his headquarters near Deep Creek. The rain, off and on most of the day, finally has stopped and the sky has begun to clear. The evening, however, promises to be cold, and Meade's fever and nausea have returned. Embarrassing, but his surgeon has nothing that will help so he must put up with it.

He's trying to get some sleep when Sheridan's courier arrives from Jetersville to the west. Sheridan's cavalry reached the village late in the afternoon with Griffin's infantry not far behind, and Sheridan's report is optimistic:

> The Rebel army is in my front...If the VI Corps can hurry up, we will have sufficient strength....I will hold my ground unless I am driven from it....My men are out of rations, and some rations should follow quickly. Please notify General Grant...P.S. The enemy are moving from Amelia Courthouse (*sic*) via Jetersville and Burke's Station to Danville. Jeff Davis passed over this railroad yesterday to Danville.

Despite his illness, the news rouses Meade to dictate a ringing order to his troops, reminding them of their past victories and asking that they now face being tired and hungry as bravely as they've faced enemy bullets.

He decides that the order reads pretty well; so well that he has a courier take a copy to Grant's headquarters. Then, leaving word that his army will take up its march at three in the morning, he takes to his cot for a few hours' rest.

LEE

At daylight Lee rides forward to join Longstreet's infantry. He's worried about Ewell's still-missing column from Richmond. Ewell's men aren't like Longstreet's and Gordon's hardened infantrymen. They're reservists, sailors without ships, and walking wounded from Richmond's military hospitals; men in uniform but accustomed to garrison life along the James, not to field duty. Others are government clerks who'd dropped their pens to take up rifles and join Ewell's march. They'd be having a hard time of it.

Still the rest of the army seems to be concentrating well and, with a little bit of luck, he'll have them together at Amelia Court House, and the 350,000 rations of bread and meat he'd ordered. Recross the winding Appomattox up ahead then five miles more to the village. Longstreet, leading the march, should be there by noon with the rest pressing hard behind.

They've been out of Petersburg's trenches thirty-six hours now. Late yesterday there were reports of men deserting and of equipment left beside the road but, as they near the village, he senses their spirit reviving.

Confederate soldiers, seeing Lee quietly waiting at Goode's bridge, cheer him, straighten their lines, and march faster.

Lee, who likes to chat with his soldiers, nods to a friendly Georgia private, "Have you been able to find any food on the march?"

"Oh, yes, sir," the boy answers, one soldier to another. "Had pretty good luck yesterday. Got a couple biscuits and a piece of raw bacon. Not a lot, but it made a fine meal. Then 'tother side of the bridge here, one of the fellers had some cornmeal and 'taters for swappin'. Soon's we can, we'll make some ash cakes. Thank you kindly for askin', General. We get to that Court House the Cap'n told us about, we'll be all right."

When most of Major General William Mahone's regiments have crossed the bridge, Lee turns to the fiery little officer, "General, we've not heard from General Ewell. Leave engineers here to hold this bridge until he crosses or they hear from me. Then destroy it."

Intent upon big problems, he takes time for smaller ones. A young Virginian, come to report that his command's across the river, is stopped short by Lee's uncharacteristically brusque, "Captain Jones, did those people (his usual term for the enemy) surprise your command this morning?"

"No, sir," the puzzled officer answers. "Had you heard that?"

"No, sir, I did not," Lee remains stern, "but something urgent must account for your appearance this morning."

When the captain still seems puzzled, Lee points: one trouser leg hangs outside a boot; the other is stuffed inside like a huge, bulky sausage.

Jones, totally crushed, can think of nothing to say. Instead he salutes as best he can and reins his horse to flee.

Lee, seeing the young man's despair, calls him back, smiles reassuringly, and speaks quietly, "Captain Jones, I didn't mean to offend you, only to remind you that all our officers, especially those who are near high commanders, must avoid anything that looks like demoralization. We must always put on our best faces. That's particularly important during a retreat. I know that you are a good soldier. Don't take my caution so much to heart, but help me."

Then it is Lee who quickly turns away. He's heard firing, up the road where Longstreet's skirmishers are pushing toward Amelia Court House. When they recrossed the winding Appomattox, the river no longer was a barrier between his army and Grant's. Now he must expect probing, slashing, cavalry attacks on his left and Union infantry closing behind him.

"What is it, General?" he returns Longstreet's salute.

"Cavalry, sir. Sheridan's. Probably Custer or Devin (Brigadier General Thomas C. Devin) feeling us out. No infantry so far, but my scouts say that Griffin's corps' over there somewhere. We can handle it."

"Yes," Lee is assured, "we're very near the Court House now. Pass your men through it, and set out security. General Gordon will guard our rear. We'll distribute rations as quickly as possible and give the men a few hours' rest. Then we must strike for the railroad. General Ewell should join us here; I've sent another courier to find him."

About 8:30 in the morning Lee rides into Amelia Court House, a sleepy little village of several unpaved streets and a score of houses half-hidden behind high, board fences softened by climbing roses and honeysuckle. At its square is an old, faded-white courthouse and, nearby, a rambling stagecoach tavern, all guarded by rows of huge oaks.

His aide, Colonel Walter Taylor, is waiting. "Sir, we've put our headquarters' tents over there. A Mrs. Smith owns the big house. She's offered it to you."

"No," Lee shakes his head. "It's very kind of her, but you 'suage her for me, Colonel. I'd rather use our tents. And we'll only be here long enough to distribute the rations. Let me see to that now."

He turns Traveller toward the railroad station and, as he nears it, gets his first inkling of the disaster the quaint little village holds for him.

There is no train puffing at the little siding. Had there been one he could have begun relaying his men toward Danville. No train. All right, so long as the rations are here. There are the boxcars, a dozen of them; and already his quartermaster general has crews opening their doors, wagons waiting to take bread and meat to his soldiers.

Then Lee hears the first anguished, frustrated cry.

"Somethin's wrong, Cap'n. There's muskets and ammunition, box after box of artillery shells and powder, harness. Lots of harness. But we can't eat that, sir. And there's no food."

"Keep looking," the officer glances sharply at Lee and orders again, "Keep looking. Hurry up! Open those other boxcars; maybe it's there."

But it isn't. There aren't 350,000 rations; there are no rations at all. Only ammunition and equipment: 96 loaded caissons, 200 crates of ammunition, 164 boxes of artillery harness, but no food. Someone has made a mistake, a terrible mistake.

Lee can't hide his anguish. The dull pain he's felt across his chest all morning becomes sharp and hard, hard enough to cause him to grip the pommel of Traveller's saddle lest the dizziness cause him to fall.

No physical blow could hurt him more than the grim situation he now faces. His army, still better than 30,000 hungry men, have marched nearly forty miles in about as many hours and, although their columns are drawn out, they're pressing hard for Amelia Court House and the food he's promised. And, when they get there, he'll have no food for them.

Mustering every ounce of the discipline the years have given him, Lee forces himself to remain calm. His face, however, can't hide his despair. For the first time since the awful third day at Gettysburg his staff see him slump in his saddle. There are great furrows along his cheeks; his ruddy complexion is suddenly pale.

He orders his quartermasters to continue searching the cars, meanwhile to send out foragers for any food they can find.

Then he sends for Longstreet and Gordon.

"See to our security, gentlemen. Be alert for any sign of General Ewell's columns. Meanwhile, I've sent our foragers into the countryside. Our people will share with us whatever they have."

"We don't dare wait long, General," Longstreet cautions. "Sheridan's on our left, maybe has Custer or Devin in front of us by now. Not enough to stop us yet; he'll need a couple corps of infantry for that. But we don't dare wait too long."

"I know, General," Lee shakes his head. "I know. But, if there's a chance of finding food here, we must risk a few hours' delay. I'll send couriers to find a telegraph still operating to Danville and Lynchburg. I'm sure that the rations are waiting just down the line somewhere, and I'll have them sent to us. Don't allow our officers and men to become demoralized. Our foragers will find some provisions."

He signs a personal appeal to be carried by those foragers:

The Army of Northern Virginia arrived here today, expecting to find plenty of provisions. But to my surprise and regret, I find not a pound of substance for man or horse. I must therefore appeal to your generosity and charity to supply as far as each one is able the wants of the brave soldiers who have battled for your liberty for four years.

His Commissary General also sends a scout, disguised as a farmer and riding an innocent-looking mule, to find an open telegraph line. The message, hidden in the scout's boot, "The army is at Amelia Courthouse (*sic*), short of provisions. Send 300,000 rations quickly to Burkeville Junction."

Lee also orders an immediate culling of equipment and horses. Horses no longer able to pull their loads will be double-teamed or left behind. Limbers, caissons, cannon, wagons, and ambulances which no longer can be pulled will be destroyed or, in the case of cannons, hidden. Artillerymen dig more "graves" on the outskirts of the little village.

All afternoon Confederate troops continue to arrive: the remains of Lieutenant General A. P. Hill's corps, the general killed in the fighting below Petersburg; Major General Cadmus Wilcox's division, those who'd survived the fight for Fort Gregg; and Major General Henry "Harry" Heth with survivors of the battle for Sutherland Station. Longstreet waves some of them into his camps west of the village; Gordon will hold the other units five miles before it.

Arriving regiments, seeing the rest of the army waiting, cheer, glad to be together again.

Then the word quickly spreads. "No rations, boys. There's been a mix-up somewhere. No rations. But Uncle Robert's working on it; he'll do something. Meantime, tighten your belts and hang on. Maybe somethin' by tonight when the foragers get back."

At sunset the forage wagons return. Almost empty. They'd not gone far beyond the village; perhaps, his quartermasters report, if they range farther in the morning, they might have luck.

About the same time a courier finally arrives from Ewell. The crippled general had found the bridges across the Appomattox carried away by high water. His engineers are laying planks across a railroad bridge, however, and with luck his footsore men should reach Amelia Court House that night.

A lantern burns late in Lee's tent that evening, and wind and rain lash the canvas as he wrestles with the choice he alone can make. At last he decides: Lee, always the gambler, will gamble one more time.

"Colonel Marshall, send our foragers out again at first light. If the enemy will allow it, we'll remain here until the wagons return or we hear where the rations are to be delivered."

A fateful decision and, even as Lee makes it, his Army of Northern Virginia has begun to break up. Gray soldiers, alone or in pairs, use the storm to quietly slip away from their camps. It's over, they've decided; we did the best we could, but now we have to look to ourselves. So thinking, they simply take up their rifles and packs and slip through the picket posts; going home.

BOYKIN

Some miles above Amelia Court House Major E. M. Boykin's 7th South Carolina Cavalry Regiment guard the rear of Ewell's long column of tired soldiers. It's after dark when they reach the Mattoax railroad bridge, now planked to allow Ewell's artillery and cavalry to cross. With the driving rain and lashing winds it's a particularly frightening crossing for men and horses. Riders gently walk skittish horses, some blindfolded, across.

Later that night Boykin's troopers cross Longstreet's picket line to camp near Amelia Court House. The rain has stopped now, and a Georgia infantryman shares with Boykin some ham and biscuits he'd picked up somewhere along the way. Encouraged with being back with the army, and having had that little bit of food, he rolls up in his saddle blanket.

The day, however, holds a final adventure for him as a campfire spreads to the hay he's gathered for his bed. Boykin wakes to find his rubber sheet and blanket on fire. As he lashes the ground with sheet and blanket, his soldiers laugh at their major's mishap. It's the best laugh they've had all day.

CLAIBORNE

Doctor John Claiborne's small medical train catches up with Confederate General William Mahone's rear guards at Goode's Bridge. Then, just beyond the Appomattox, Mahone himself waves to Claiborne from the yard of a small plantation.

"Come on, Doctor, have a bite with us."

Claiborne joins Mahone's staff as they eat chicken and biscuits and chat as if they hadn't a care in the world. When they've finished their meal, however, Mahone hurries them back to their saddles.

"Time we were off, Doctor," the general explains and points across a wide field to a long line of advancing infantrymen.

"Are those *Yankees*, General?" the incredulous surgeon asks.

"Reckon so. The last of my men passed by us about a half hour ago."

"So there's no one between us and the enemy?"

"'Pears like it, Doctor," Mahone grins. "That's why it's time we were off. Say, if you're not under orders, would you like to attach yourself to my staff?"

Claiborne, not sure that he can handle the fiesty little general's casual approach to Yankee soldiers, asks, "General, you have a very good surgeon on your staff don't you?"

"Yes, Doctor Wood. A good man."

"Then, sir, I think that I may be needed more at General Lee's headquarters."

"As you wish," Mahone grins again, "but you'll be sorry you didn't stay with us."

DAVIS

At Danville, a hundred miles southwest of Amelia Court House, Jefferson Davis anxiously waits for some word from Lee.

He inspects Brigadier General-Rear Admiral Raphael Semmes' work on Danville's breastworks then retires to his temporary office to draft a proclamation calling all Southerners to rally against the Northern invader.

Not many Southerners will ever see it, and those who do must wonder at the soundness of its argument:

> We have now entered upon a new phase of the struggle...relieved from the necessity of guarding cities...with our army free to move from point to point...and where the foe will be far removed from his own base...nothing is now needed to render our triumph certain, but our own unquenchable resolve....

In the evening, when Davis still has not heard from Lee, the Danville telegrapher clicks an anxious query up the Richmond-Danville Railroad line. Will anyone volunteer to establish contact between the Confederate government in Danville and Lee's army?

JOHN WISE

Cavalry Lieutenant John Wise, still on duty at little Clover Station, reads Davis' wire and at once volunteers. An engine with tender and baggage car are sent rattling up the tracks to him. Wise is to try to get as far north as Burkeville. If he makes it there, his locomotive is to switch over to the Lynchburg line of the Southside Railroad then ride west. Lee, they guess, will try to swing west then south, to Danville; if he can't do that, he'll probably march for Lynchburg, staying close to the railroad for supplies. If the railroad is blocked below Burkeville, Wise is to send the train back and attempt to reach Lee overland.

The little train, with Wise riding alone in the baggage car, sets out from Clover Station. As it passes one abandoned station after another, the young officer strains to see the passing countryside, impossible in the cold, foreboding darkness.

Finally, unable to see and feeling helpless in the baggage car, he makes his way to the locomotive. He and the engineer then decide to abandon the baggage car and continue north, as fast as they dare go without a headlight, until they can go no farther.

About 2 a.m. they reach deserted Meherrin Station, below Burkeville. While the engineer loads his tender with wood and water, Wise runs to bang on the door of a nearby house. Its sleepy owner, lantern in hand, finally appears.

"Have you heard anything of Lee's army?" Wise asks.

"Naw, nothin' at all," the man answers, "'cept last we knew he's supposed to have been at Amelia Court House."

A few more questions, unanswered, then Wise asks, "How about the Yankees? Any idea how close they are?"

"God knows," his informant scratches his head thoughtfully then adds, "I reckon that today or tomorrow they'll be everywhere. They say that up the line here it's already crawlin' with Bluebellies. Reckon you'd better make yourself as scarce as the rest of our army."

Not able to do that, Wise climbs back to the locomotive's cab and persuades an increasingly nervous engineer to race up the tracks toward Burkeville.

At last, as they near a long curve, the engineer slows. Over the dim track before them, Wise sees a glow on the horizon but no sign of their enemy.

"Burkeville?"

"Just 'round the bend, Lieutenant; but I don't like all that light up there. Station's usually deserted this time of night. Best keep your eyes peeled."

Slowly they ease around the curve, the engine's headlight still unlit. Several hundred yards ahead Wise sees Burkeville Station.

Huge fires are burning beside the tracks. Outlined in the flames' garish light are a dozen teams of trackmen. Despite the cold, some are stripped to the waist, pounding railroad spikes or carrying long rails. Some of the workers wear dark uniforms. It's the young officer's first personal view of the enemy, living, breathing, capable of killing. He's alarmed but strangely exhilerated and relieved. At least the long waiting is over.

They're laying track, changing the gauge of the Southern railroad to accommodate the wider gauge of Grant's Military Railway. No sign of armed guards, though there must be some nearby, and Wise reasons that means that they feel secure as they work. Grant's army must have passed beyond Burkeville; Lee then must be somewhere north or west.

The track crews are so intent on their work that for several minutes no one notices the single, darkened locomotive quietly puffing several hundreds yards down the Danville track. Then someone shouts, points, and the workers scatter. Some of them, recovering quickly, grab weapons left by the station's platform, and begin to run toward Wise's locomotive.

"Reverse the engine! Back her up!" Wise calls to the engineer. The trainman, however, either too frightened to move or too sensible to try to outrun rifle bullets, hesitates.

Drawing his revolver, Wise again orders, "Back her up, Mister, or you're a dead man! Either they'll do it or I will. Back her up!"

Slowly at first then much faster the locomotive and tender puff back around the curve then, picking up speed, rattle down the dark track. Their pursuers soon give up the chase; but Davis' couriers run at near full throttle until tiny Meherrin Station again appears.

As the train slows, Wise yells to the engineer, "Go back to Danville. Tell them what's happened and that I'm going ahead on foot to find Lee's army. I'll report when I can."

Then he jumps from the train, tumbling and rolling along the gravel track. Momentarily stunned, he staggers to his feet to catch a last glimpse of the locomotive disappearing below the station.

Feeling very alone, he checks the loads in his revolver, recovers his hat, dusts dirt from his uniform, and sets out.

He's walked several miles up a dark country road when he comes to a lone farmhouse. Tied to a rail before it is a pretty mare, saddled and bridled; a godsend.

A lantern glows in a kitchen window and, in answer to his knock, a Virginia cavalry trooper, faded sergeant's stripes on his sleeve and a large Colt revolver in his hand, appears at the door.

––––––––––––––––––

"I'm sorry to bother you, Sergeant," Wise glances again at the man's uniform, "Lieutenant John Wise. I've been at Clover Station, but I'm on a special mission for President Davis. He's in Danville; sent me to find General Lee's headquarters, somewhere above us. Have you heard where our army is?"

"Now, wait a minute, Lieutenant," the soldier speaks softly, but Wise is uncomfortably aware that, although the man is half-smiling and is talking quietly, as if they were discussing the weather, the revolver in the sergeant's hand is pointing at his belt buckle and it's not wavering an inch.

"Wait a minute, Lieutenant," he repeats. "You come in here, afoot, early in the mornin', askin' questions about the army. Far as I know, you could be a Yankee soldier. Now you'd best show me some proof that you're what you say. You'd best show me pretty quick or I'm gonna get awful nervous."

"No, wait," the quickly aging lieutenant fumbles at his jacket pocket, "here's my orders. A telegram from President Davis at Danville."

The sergeant holds the telegram before the lantern's light, his revolver never moving from Wise's belt buckle. Meanwhile, Wise continues to explain what he knows of the fighting for Richmond and Petersburg and Lee's retreat.

Finally, the sergeant, convinced, holsters his revolver and returns Davis' telegram.

"All right, Lieutenant. I'm on medical leave; plan to find the army myself in a few more days. How can I help?"

"I need to borrow your horse. The mare I see tied up over there. I'm afoot, and I've got to find Lee's headquarters then get back as quick as I can. A day, maybe two."

A pause then, "All right, I believe you. I'd go along, but I'm not up to settin' a horse yet, and this is the only one I have anyway. You take the mare. But I want her back in a couple days. Understood?"

"Yes; you'll have her. Now any suggestions on how I can find the army?"

"You saw the Bluebellies at Burkeville. Their army could be anywhere from there, but I'd head on up this road. Then swing wide and come in on Farmville from the west. If the timin' is about right, you should find them there or at least some word of them."

"Thanks, Sergeant," a relieved Lieutenant John Wise shakes his hand and turns to run to the horse.

As he springs into its saddle the Confederate sergeant reminds him again, "Mind you take good care of June there, Lieutenant; she's a mighty good horse."

Wise grins, waves good-bye, and gallops up the dark road.

As he rides, it occurs to him how quickly his prayer that he be relieved of the boredom of guarding deserted Clover Station has been answered.

"Careful, Lieutenant," he scolds himself, "sometimes you get exactly what you ask for."

PEMBER

Back in Richmond, Phoebe Yates Pember has absorbed a dozen wounded Yankee soldiers into her ward of wounded Copnfederates and fought off a band of deserters only to find that her greatest concern, one she thought resolved, is not resolved at all.

The promised rations have not arrived, and the hospital's commissary, now under Yankee management, refuses to issue her anything.

"Our stock's gone, Miss Phoebe," Jim, her ward boy, reminds her.

"I gives these gentlemen all we has this morning. Not worth a shucks as it was, but there's nothin' at all for tonight. And the Yankee soldiers done read your note and laughed. Said we ain't got 'proper arboration', or somethin' like that. Hunh, as if they knowed what they's talkin' 'bout. Why you's practically a doctor yourself. What we goin' to do, Miss Phoebe?"

"Do? Do? Why we're going to give them a chance to correct their mistake, Jim. You tend the ward; I'll be gone for awhile."

In her room she selects what she calls her "full-dress Confederate uniform": boots of untanned leather, tied with thongs; a homespun dress of black and white blocks, the white: cotton yarn, the black: an old silk, scraped into pulp with broken glass then respun; cuffs and collar of bleached homespun; a hat plaited of straw taken from the fields back of Chimborazo, dyed black with walnut juice and adorned with a shoe-string ribbon; and knitted gloves dyed in three distinctly different shades of green.

Thus splendidly equipped and armed to the teeth with righteous indignation, Phoebe Yates Pember pushes aside the bayoneted rifles of the Yankee sentries at the commissary's door to accost its Officer in Charge. "I am Mrs. Phoebe Yates Pember, Captain. Matron of Chimborazo's only remaining ward. I have two dozen wounded Confederate soldiers and a dozen of your own men under my care. I was promised provisions, to be delivered this morning, but I've not seen them. Unless you intend to starve your own men, as well as captured sick, you must issue food to me now."

"Mrs. Pember. Of course; I've heard of you." He smiles disarmingly and adds, "We've all heard of you. I'm sorry, ma'am, but our supply boats haven't been able to get through the obstructions in the river."

"I'm not surprised, sir. Our sailors are quite efficient. But then you must return my horse and carriage, which your soldiers took last night. I have some coffee I've saved for months from our rations. I can trade it and a small amount of whiskey for food."

"Why were those supplies not turned in, ma'am?"

"Captain," she patiently explains, "surely you've been around the army, any army, long enough to know that you just do not surrender supplies of that nature. They're good for bartering, sir, and I mean to barter. Now, the horse and carriage, please."

She's intelligent, pretty, and angry. Hard to resist.

"Are you a Virginian, ma'am," he attempts to disarm her as he weighs the chances of his colonel's finding out whatever he does.

"No. I am a South Carolinian, come to Virginia at the start of this war to try to help ease the suffering in our hospitals."

"I lost a brother in South Carolina."

"I'm sorry. So many lives have been lost on both sides."

"Yes," he continues, "and I see in the pale faces and pinched features of Richmond women how much you too have suffered in this war."

"If our features are pale and pinched, sir," she flares, "it's not because we lacked under our Confederacy but because we sorrow at its failure."

She instantly regrets her words; he means well. The Yankee captain, however, seems more amused than angered by her answer.

"I meant no harm, ma'am," he smiles. "You'll have your horse and carriage; and perhaps we can find a few supplies here to tide you over until the river is free."

She loads her carriage with the food and medical supplies she graciously allows him to persuade her to accept, more than the Confederate commissary would have issued her in a week. Then, loading a small bag of coffee and a demijohn of whiskey in her carriage, she drives to Richmond's market where she trades both for a live calf and other provisions.

When she returns to the hospital she finds the Vermonter sentry at its door already admiringly aware of her adventures with the Yankee commissary officer.

Leaning his bayoneted rifle against the door, he carries her treasures to the pantry. Then he ties the calf to the back of the building.

As he works he compliments her as sincerely as had her Confederate patients.

"Ma'am you sure do beat all! I worked up in Burlington a couple years. Then I was on a lake steamer crossin' Champlain before I come into the army. Saw a lot of fine ladies. But there weren't a one of them could hold a candle to you."

When he's finished helping her, he politely refuses the drink of whiskey her gratitude offers him. Then he takes up his rifle at the hospital's door, but reminds her to call him any time she needs help.

Late in the afternoon she returns to the commissary for sugar.

"Sorry, ma'am," one of two new Yankee soldiers on guard there politely tells her, "the clerk's not here."

"That's all right," she answers, and they look on in astonishment as she brushes by their bayonets, uses the key she'd taken from the captain's office, and proceeds to find two bags of sugar. Then she has one of them carry them to her carriage.

That done, she re-locks the door, bestows on the sentries her best smile, and tells them, "If my coming and going cause you gentlemen any problem, and my arrest be necesssary, please tell your captain that I shall be at my ward."

It isn't necessary and, from that moment on, she'll have free run of the Federal storehouse. No explanation is ever given her except Jim tells her that he heard her Vermont soldier say that the Yankees are "awful skeered of her."

LINCOLN

At City Point Lincoln also has risen early, breakfasted, read dispatches in the telegraph office, and now prods Admiral David Porter.

"Come on, Admiral. According to General Weitzel, the fires are out; the riots ended; let's go see Richmond. For four years I've been living a horrible nightmare, and now it has ended. Thank God I've lived to see it! Let's go see Richmond."

There's a lot of risk to the trip, a lot of risk. Seeing Lincoln's determination, however, Porter gives in.

They start aboard the *Malvern*: Abraham Lincoln, Tad Lincoln, Admiral Porter, and William H. Crook. Led by Captain John Barnes on the *Bat*, with an escort of sixty marines, and preceded by a small flotilla of naval vessels to clear the James of mines and other obstacles.

By noon they've cleared Dutch Gap, the farthest Union outpost upstream, and Porter warns his lookouts to be especially alert: they're on a dangerous stretch of the James.

On the river's banks, he sees elaborate Confederate earthworks and artillery revetments, abandoned now but offering hundreds of possible hiding places for any sharpshooters left behind to delay Union ships' advancing on Richmond. And the water has its own hazards: its channel littered with charred timbers, the swollen bodies of horses, and the masts and pilothouses of vessels deck-deep in the James River's muddy bottom. On its banks, the presence of dozens of beached mines, already dragged from the river, warn that there may be many more explosive devices hidden beneath its surface.

Lincoln, however, enjoys the clear, warm spring morning. Anticipating seeing Richmond, perhaps just around the next bend, he seems unaware or unconcerned about the risk.

Past Chaffin's Bluff the James becomes too restricted for the *Malvern* or the *Bat* so an uncomfortable Porter transfers his passengers and two squads of marines to a small barge. Then, towed by the navy tug *Glance*, they continue upstream.

They've not gone far when they come upon another obstruction. The former New York City ferry turned gunboat, *Commodore Perry*, one of many navy ships hoping to be first into Richmond, has run aground. As they near the beached ship, the *Commodore Perry*'s captain, unaware of Porter's approaching barge, is furiously spinning his paddle wheels to free his ship. Mud and water fly everywhere, and Porter's barge is nearly swamped before a lookout on the *Commodore Perry* sees the *Glance* and its barge and shouts an alarm. Then the *Commodore Perry*'s engines stop.

Porter, purple with rage and not needing the bull horn an ensign hands him, calls for the *Commodore Perry*'s captain.

"Captain Amos Foster, sir," the captain answers.

"Admiral Porter, Captain," he roars. "You nearly swamped us, and I have the President aboard. If your lookouts are blind, I suggest you replace them. Understood?"

"Yes, sir. My apologies."

"Can you get yourself off that mud bank?"

"Aye, sir; just a matter of time."

"Well, when you do back off it," Porter roars again, "don't attempt to go on to Richmond. Drop your anchor and let the other boats go by. Understood?"

"Aye, sir."

Their tug, *Glance*, swings clear of the beached *Commodore Perry* and continues upriver. When it's about a mile below Richmond's landing, however, they encounter still another Federal ship, the *William Allison*, also aground. Then the *Glance*, attempting to free the *William Allison*, becomes stuck too.

Porter, nearly apoplectic, orders his marines to man the barge's oars. The presidential party, reduced now from a score of naval vessels to a single barge rowed by a score of marines, slowly makes its way to Rockett's Landing.

Lincoln, who has watched the steady shrinking of Porter's fleet, chuckles.

"You know, Admiral Porter, I'm 'minded of a fellow who came to me some months ago asking to be appointed an ambassador. When I told him, 'Well, I'm sorry but that just wouldn't wash with the people,' he lowered his sights a bit."

"Then," Lincoln continues, "he asked for a more modest post. I turned him down for that one too."

"Well, he thought a moment then asked, 'How 'bout makin' me a tide-waiter?'"

"'No,' I told him, 'can't do that either.' Well," Lincoln chuckles, "he kinda sighed; then he asked, 'reckon you got an old pair of pants I might borrow?' It's well," he concludes, "to be humble, Admiral; well to be humble."

Porter laughs, but he's becoming more and more concerned about their safety. At Rocketts Landing before them he sees no Federal warships nor Union troops. Smoke still hangs low over the city; and he has no way of knowing whether Weitzel really has everything under control. No way either of knowing how Richmond's citizens will react to Lincoln. And Porter has only a handful of marine guards to protect the President.

Then their barge reaches the dock, and they clamber out and look around. Scanning his charges, Porter must admit that they're a strange-looking group. Lincoln in black suit, white shirt and string tie, stove-pipe hat, and prairie brogans, patiently waiting but obviously determined to enter the city. So tall he can't be mistaken. Crook, his civilian guard, holding Tad Lincoln's hand while his other hand is hooked in his belt, near a large revolver. Short, stocky Porter on Lincoln's other side. Marines in short jackets and baggy trousers, apprehensively clutching their weapons. Not exactly the image of a conqueror but, Porter smiles grimly, Lincoln wouldn't have it otherwise anyway.

They've barely reached the end of the landing when Lincoln, recognizing a Northern reporter, Charles Coffin, standing nearby, calls, "Mr. Coffin, good to see you here. Do you know where General Weitzel is?"

"Yes, sir. Come with me."

Before they can follow, however, Coffin, sensing a story, turns to several Negro laborers to ask, "Would you like to see the man who set you free—Abraham Lincoln? There he is."

Then Lincoln's party is quickly surrounded by a dozen jubilant Negroes. One, a white-haired older man, grasps Lincoln's hand before the President's guards can stop him, pumping it and shouting, "Bless the Lord, the great Messiah! I knowed him as soon as I seed him! Hallelujah, Lord Jesus! Come at last to free his children from their bondage! Hallelujah!"

Lincoln, embarrassed, tells him, "No; not me. Thank God for your freedom for He has given us this victory. But so long as I'm alive, you'll not wear chains again."

More and more Negroes have arrived, and suddenly they're all singing a hymn, "All Ye People, Clap Your Hands," and joining the presidential party in an impromptu parade up the narrow cobblestoned streets into Richmond.

They pass near porches, balconies, and steps crowded with people watching. Some call out a welcome; others are sullen, quiet, unresponsive.

Porter and Crook are increasingly alarmed. Once Crook sees a man standing in a window, holding something long and slender in his hands.

"Careful, Admiral; over there!" and Crook draws his revolver.

Then the figure disappears; the window is clear.

"Move on," Porter waves his arms, and Lincoln's party plod on. The President, his face calm, seems not to have noticed the danger, and Porter realizes again that Lincoln is determined to follow his quest to the end, accepting whatever happens.

They pass the one-time tobacco warehouse turned into Libby Prison, for many months crammed with Union officer-prisoners. A sinister-looking place, even in its deserted state. The crowd of Negroes around them, seeing Lincoln's face harden, yell, "We'll pull it down, sir; we'll pull it down!"

"No," Lincoln waves his arm. "Let it stand; let it stand to remind us."

A young white woman stands near the Exchange Hotel, an American flag draped over her shoulders, waving. She's the only one, Porter realizes, who seems friendly. Another mute witness to the procession, a Southern matron, later will tell her grandchildren, "I saw Abraham Lincoln as he entered our city that day. And, with due respect to the President of the United States, I felt him the ugliest man I'd ever seen."

A pretty girl pushes through the crowd to hand Lincoln a bouquet of roses.

A rough-dressed man follows, trying to force his way to Lincoln's side. Abruptly halted by a marine bayonet, the man snatches his hat from his head and explains, "I mean no harm, only to shake the President's hand."

Again, Lincoln pauses to touch a baby held by its Negro mother. "Thank you, dear Jesus," she holds up the child; "thank you."

Lincoln walks along, the slow, steady pace of a Midwestern farmer; past the governor's mansion then three blocks more to Weitzel's headquarters, the former Confederate White House. Sweaty and tired from his two-mile walk, and now joined by Colonel Ripley ("General Weitzel's been notified and is on the way, sir"), in charge of the Federal troops occupying the city, he enters the home Jefferson Davis had vacated only two days before.

Seeing Ripley and a large contingent of Union soldiers, Admiral Porter, for the first time that day, relaxes a bit. So does Lincoln.

Looking tired or perhaps just deep in thought, he crosses the cream-colored rug in Davis' office to sit at the large desk. The President doesn't share his thoughts at that moment but Porter, watching him closely, realizes that the tall, gaunt, haggard-looking man must be thinking of all that has passed since he became President of the United States more than four years ago. So much suffering and heartache and worry. More than 600,000 Americans already dead in their long civil war, and it still not ended. And now, decides Porter, after all the bloodshed he must feel that at last, for the first time, he really is the President of *all* the United States.

Lincoln ends his silence to ask if he might have a glass of water. Then, "Is the housekeeper gone?"

"Yes, sir."

"Well," suddenly recovered, he jumps up like an excited boy, "let's look at the house."

Weitzel's aide, Captain Graves, escorts them through Davis' home. On its back porch Lincoln thoughtfully pauses.

"This must be the porch from which Mr. Davis' son, Joe, fell to his death. So sad. I lost my son too, you know. So sad."

Then he's called back to Davis' former office. General Weitzel and two Confederate officials: Judge John A. Campbell, one of the Confederate peace commissioners several months ago; and General Joseph Anderson, the Tredegar Foundry director, have arrived. Other state and city officials, learning of Lincoln's presence, are on the way. All of them want to see him, and Porter notices that the boyish expression he'd enjoyed on Lincoln's face now is gone. There's work to be done, and he can't turn from it.

Lincoln meets privately with Judge Campbell.

"I have no authority to meet with you, Mr. President," Campbell begins, "but I too want this war to end. Two things are necessary. First, we Southerners must order our armies to lay down their arms. Then our state governments must vote their states back into the Union. My concern, as a Virginian, is that we lead the way."

"What do you have in mind, Judge?" Lincoln thoughtfully nods.

Campbell uses his fingers to count the names of a half-dozen prominent Virginians, including General Robert E. Lee. "You authorize our meeting, Mr. President, and I'll summon them here. Then we'll vote Virginia out of this war. With Virginia out, the other Confederate states will follow."

A pause then, "I appreciate your intentions, Judge," Lincoln thought-fully answers, "but I'm not sure that's the best way to go. Let me think about it, and tomorrow we'll meet again."

When Lincoln suggests remaining in Richmond overnight, however, Porter won't hear of it. Even as the admiral rejects the idea they're inter-rupted by cannon fire from the river. To their relief it's Porter's flagship, the *Malvern*. Its captain, determined to regain Porter's favor, has found his way up the James then announced his arrival with a cannon salute.

The *Malvern* will do, Porter decides. But not docked; anchored in mid-stream. And with extra marine guards aboard.

They return to the *Malvern* in a cavalry-escorted carriage. On the way they again pass Libby and Castle Thunder prisons. Noting Lincoln's thoughtful expression, Weitzel asks, "Mr. President, have you any sugges-tions on how I should deal with Richmond's citizens?"

Lincoln smiles gently, "If I were you, General, I'd let 'em up easy."

Before his dinner on the *Malvern*, Lincoln has a chance to demon-strate his "let 'em up easy" philosophy, but it's with a Yankee sailor.

As they returned to the *Malvern*, Porter spotted the *Commodore Perry* anchored nearby and exploded, "I told that captain—Foster, yes, Foster—not to come on to Richmond. Now he's disobeyed my order. I'll see to that."

Soon after, he's summoned Foster to his flagship.

"Cap'n Foster of the *Commodore Perry*, Admiral," the officer salutes.

"I know who you are, Captain. And I know that I told you that when you got free of that mudbank you were not to come onto Richmond but to give way for other ships. You have disobeyed me, sir! Explain yourself."

"Sir, beggin' the Admiral's pardon," Foster quietly corrects him, "but that's not the way I remember the Admiral's orders. You said that when I *backed* her off the mudbank I was to give way to the other ships."

"Yes, yes," Porter interrupts the explanation. "That's what I said."

"Well, sir, I didn't *back* her off. I gave the old girl everything she had and drove her *over* the mudbank. As we didn't *back off*, sir, it seemed to me that your order about Richmond didn't apply."

Having stated his case, Captain Amos Foster stands patiently at at-tention while Porter, trying hard to suppress a smile, turns to Lincoln.

"Mr. President, as a lawyer what would you advise in this case?"

Lincoln chuckles, "I'm just a prairie lawyer, Admiral. Don't know a heap about maritime law. But it seems to me that Captain Foster here sort of has you over a barrel. Let him up easy, Admiral; let him up easy."

"Were you," Porter turns again to Captain Foster, "the first naval vessel to reach Richmond, Captain?"

"Oh, no, sir; the Admiral's boat got here just a hair ahead of us. As it should be, sir."

"That's right, and don't you forget it. All right," Porter grins, "go on; get out of here; and tell that crew of yours that their Commander-in-Chief saved every last one of you from being keel-hauled."

Still later, when it's dark and all is quiet on the *Malvern*, Porter has two final scares to end a very trying day.

First there is a hail from the dark shore, "Ho, *Malvern*! Dispatches for the President. Send a boat!"

Porter is cautious. "Send some marines to pick up the message, Captain, but don't allow the messenger on board. Keep him under guard at the dock."

The marine officer soon returns, "Sir, he won't give up the dispatch. Says he must deliver it in person."

"Very well. Search him thoroughly. Then bring him on board, but watch him very carefully. If he makes one false move, kill him."

A few minutes later the captain again returns. "Sorry, sir. When we went back he'd disappeared. No trace of him."

Soon after, the silence again is broken by a call from the shore: "Ahoy, *Malvern*! I'm a sailor off the *Saugus*. Let me come on board."

Porter again is alert. "There's no *Saugus* in the fleet, Captain. Take your marines ashore and arrest that man."

A bit later the marines report that this man also has disappeared.

"Very well, Captain. Place two of your men outside the President's stateroom. Keep them there so long as he is aboard. Allow no one near him."

Lincoln, uneasy over the incidents, thanks Porter for his concern and turns in for the night. It's been a very busy day, a full day in so many ways. He's sure that morning will bring another; he'll need the rest.

SHERIDAN

General Philip Sheridan, also up at dawn, pushes west all day in the overcast, rain, and wind.

He's split his cavalry into two columns: Major General George Crook leads one to cut the railroad ahead of Lee; the other, Devin's and Custer's men, slashes at Lee's drawn out column near Amelia Court House. Meanwhile, Griffin's V Corps' infantry march hard behind Sheridan, striking for Jetersville, some six miles below Amelia Court House on the Richmond-Danville Railroad.

Devin and Custer hope to force Lee to turn and fight while Crook blocks his retreat somewhere between Jetersville and Burkeville. Whichever scheme works, once Lee turns to fight, Griffin's infantry will be needed.

Griffin's regiments begin their march before dawn, Brigadier General Joshua L. Chamberlain, Griffin's hardest fighting general, leading their advance. And once more, to the dismay of Sergeant Thomas McDermott, Chamberlain's faithful orderly, the bruised and hurting Chamberlain, still recovering from the fresh wound he'd taken in the Quaker Road fight just a few days before, insists on walking beside his men.

Griffin, hearing his staff officers chuckling at the sergeant's pestering his general to ride the horse, recalls hearing McDermott, early that morning, in his most respectful voice, ask, "General, darlin', if you don't mind too much, won't you just once remember you're supposed to ride the damned horse, not lead it?"

Griffin's staff, having heard the sergeant's argument so often that they've taken to calling Chamberlain's horse "damned horse," are chuckling as they ponder Chamberlain's persistence in walking.

"He knows what he's doing," Griffin interrupts their discussion. "You can bet that his men know how badly he was wounded before Petersburg last summer. Everyone thought he'd die; and he read his own obituary in the New York paper. Then he walked out of the hospital to come back to this corps because he knew we'd soon be in a fight. Then, two days ago, they saw him wounded again: dirty, his blouse all torn, covered with blood, but right up there leading the fight. Gentlemen, if Joshua Chamberlain not only marches but sets the pace for everyone else, no one's gonna hold back. Check the straggler line. I have, and I couldn't find a single one of his men there."

Griffin does his part to move them along, calling out to soldiers sitting beside the road, "You men! What's the matter? You fellows are dropping back."

"Clean wore out, General," one answers. "Don't reckon we can march another step."

"Now look here, boys," Griffin reasons with them, "we finally have Johnny on the run; and the old V Corps' fixin' to head 'em off. If they escape us, Sherman's bummers will catch 'em and get all the glory. And we won't have a thing to show for four years' fighting! Come on now; left, right, left, right. Say it to yourselves. That's it. First thing you know you'll be back with your regiments. Come on now; don't flicker out at the last!"

As they march they see Confederate stragglers, alone or in pairs, slipping through fields around them. If the stragglers offer no threat, they ignore them.

"Never mind the Johnnies," they argue. "They're all wore out and fixin' to go home. Had enough. Let 'em go."

At one point Chamberlain's skirmishers see several Confederate officers ahead of them, leading Gray soldiers into a nearby woods. Then one of the Confederate officers signals the Union soldiers to hold their fire and, when they motion to him, rides over to explain.

"Don't shoot, boys. We're Major Young's scouts. The major's down the road with a whole passle of Johnnies, and we're takin' him some more."

Sure enough, a mile beyond they come across Major Henry Young, quietly sitting beside the road, his horse grazing nearby, while he watches his scouts herd more Confederate soldiers into an improvised pen.

Major Henry Young's scouts also bring in a lean, hungry, ragged would-be farmer on a decrepit old mule. They'd picked him up near Jetersville. His story sounded reasonable enough, but they'd searched him anyway. In his boot were two identical messages, one addressed to the Confederate commissary officer at Danville, the other to his counterpart in Lynchburg:

The army is at Amelia Courthouse, short of provisions. Send 300,000 rations quickly to Burkeville Junction.

"Good work, Young," Sheridan beams. "Now, maybe you ought to send these messages on for General Lee. If they take the bait, you might catch us some extra rations."

Young grins and is gone. The Confederate commissary at Danville isn't fooled; the telegram to Lynchburg, however, causes a train to be loaded with provisions and dispatched for Burkeville. Young and his scouts plan to welcome it.

Ahead of Griffin's infantry, late in the afternoon Sheridan's cavalry reach Jetersville, a nondescript little village clustered around a small railroad station: a dozen weather-beaten homes, a store or two, a blacksmith shop, and a post office.

"Glory be!" their commander shouts. "No Johnnies. Reckon we're ahead of them!"

They picket their horses and work fast to throw breastworks across the road leading toward Amelia Court House. Knowing that Longstreet's men must be up that road speeds their work.

They talk as they dig, reasoning among themselves.

"We're ahead of 'em. That's the long and short of it. But where in hell's the V Corps? Damned if that Chamberlain fellow don't march 'em hard. Foot cavalry, gotta give 'em that. Well, they'd sure better march today. Can't stop Johnny without infantry. Never thought I'd be glad to see infantry, but I reckon it's so."

As the light begins to fade, Sheridan's troopers hear scattered rifle fire somewhere up the Court House road. Never mind; keep working. They stop now and then, however, to catch their breaths and to stare up the dark road or to glance back over their shoulders and mutter again, "Where in hell's the V Corps?"

As Chamberlain's skirmishers reach the little village, the usual jibing between cavalry and infantry begins.

Sheridan's cavalrymen, suddenly no longer apprehensive, start it.

"Hey, Billy, where you been all day? Thought you'd never get here." Or, more admiring, "Where'd you hide your horse, Billy? Must've rode him nigh to death this day. Thirty-five miles you say? Hallelujah, that's a fur piece."

Chamberlain's tired infantrymen grin at the teasing and give as good as they get.

"Had to get here 'afore you fellas cleaned out every henhouse in sight." Or, "You fellas can git on home now; we'll take care of things here."

They've come a long way this day. A cold, wet, hungry, muddy, foot-slogging infantry march that nearly kills Chamberlain and a great many others, but Griffin's corps is *up*; *up* beside Sheridan's cavalry, *up* before Lee's skirmishers. At Jetersville; on the critical Richmond-Danville Railroad.

Despite the hard march, they feel pretty good about things. They're out of the awful trenches and on the road again. And this time the end really *is* in sight. So many times they've been the ones retreating; now the shoe's on the other foot.

"Maybe," they reason over their campfires, "if they come on tonight or first thing in the morning, we won't be able to stop 'em, but we'll sure give 'em what for. Walked too far, too fast not to make it count now."

At last Sheridan turns in for the night. It will be another short one for him. Lee probably won't attack now until morning, he reasons, and, unless Meade has another infantry corps up by then, it's apt to get pretty hot around here, very early in the day.

CHAPTER 3

Precious Time Lost

April 5–6, 1865

LEE

He sits in his camp chair, the worn quilt that has seen him through so many campaigns draped around his legs. The night has been short, and he's spent it there in the chair. When he isn't dozing, he seems to think better sitting up that way, weighing the few options remaining. The dull pain across his chest and shoulders eases when he's sitting.

Rain, a hard-driving, cold rain, still lashes the canvas. He's grateful for the tent, but his mind constantly drifts to his soldiers, most without any shelter at all. He thinks of Ewell's men marching in the night before and wonders how they're faring in the storm.

He smiles, recalling Commodore John Randolph Tucker's improvised "Naval Brigade," still clinging to their extra uniforms, pots and pans, one even clinging to a precious accordion though he looked tired to death from the long day's march.

Feeling that Tucker's men couldn't keep up anyway, Ewell relegated them to his rear guard. Somehow, however, the sailors and their commodore *had* managed to keep up fairly well. Maybe knowing that they were *the* rear guard had been an extra incentive. And their commander, Commodore Tucker, wasn't about to be shown up by any army people. Tucker, far more accustomed to pacing a quarterdeck on a warship than to a long, dusty infantry march, calling out orders strange to listening soldiers but customary to his grounded sailors: "To the starboard, march!"; "Hoist your sails, Mr. Brown; we've a long way to port this day"; "A little farther, men, and we'll drop anchor for the night."

51

When Tucker sailed his brigade through their picket line, Longstreet's infantrymen laughed at the sailors' uniforms so far inland and at the commodore's strange commands. The more thoughtful veterans cheered, however, knowing that Tucker's sailors had marched even farther than they that day. And General Kershaw spoke of them with conviction.

"Never mind their appearance, General Lee. Credit them with the day's work they've done. By morning they'll have discarded some more baggage and, if that's not enough, pride will keep them marching. They'll keep up, sir. And General, when the time comes, they'll fight. That Tucker," he chuckled, "will see to that. You know, our infantry's used to turning back an assault before it gets down to the bayonet. These men are trained to exchange broadsides at thirty yards then to settling the matter with cutlasses on a narrow deck where there's no chance to run or hide. They'll do, General; they'll do."

Lee nodded, then asked, "And the others?"

Ewell, his pinched features etched with fatigue and the pain of his ill-fitting wooden leg, answered.

"Bureau clerks, General. Quartermasters, commissary officers, ordnance men. Good enough in their element, I suppose. Fine clothes; but they'd best not take their boots off near Longstreet's boys or, in the morning, they'll be barefoot. Most of them don't have weapons, and those who do have never fired them. And you can see the fear in their faces, General. Never smelled burnt powder, never faced a man with a bayonet. They kept up today, but we'll have to watch them. They'll skedaddle for home the first chance they get. More a hindrance than a help."

"Well," Lee smiled, "They're what we have, General, and they are *yours*; you must make the best of it. Share the food you have; bed them down. In the morning organize them into companies and regiments. I don't know how much time you'll have. It depends upon our foragers and what our enemy will allow us."

Then he smiled again, "You both did well. I was concerned about you. Now that our army is together again, I expect that things will look brighter in the morning."

———————

Near morning now; he hears Colonel Charles Marshall quietly raising the tent flap. If Marshall thinks he's sleeping, he'll allow him a little more rest.

Lee speaks softly, "It's all right, Colonel Marshall. I've been awake for some time. Any word of our foragers?"

A pause then Marshall's reply, "Most are back, General."

Lee smiles reassuringly, "It's not like you to leave a report incomplete, Colonel; have they found provisions for our army?"

"Very little, sir. The people want to help, but they have nothing to give." Then he adds, "I'm sorry."

Lee nods, "I feared it, but we had to try. Now we must look to other alternatives. The enemy, they are quiet?"

"Yes, sir. Generals Longstreet and Gordon both report light probing."

"No word of rations from Danville or Lynchburg?"

"Not yet, sir."

"Then we must look to ourselves. I'll be ready in a few moments. Have Sergeant Tucker ready to guide us to General Longstreet's headquarters."

By mid-morning he's issued his orders for the day's march. Longstreet's two divisions will lead. They'll be followed by Anderson's and Pickett's men. Then Ewell, with Gordon protecting their rear. Fitz Lee's cavalry will guard their flanks and probe down the road to Jetersville then to Burkeville and the rail line. Burn whatever our horses can't pull. Officers must keep the long column closed up on the march.

Maintaining discipline is very important. During the night, he's told, fifteen men of the Richmond Howitzers deserted, along with a whole company of the 9th Virginia Infantry Regiment. Commanders report straggling. On the village's lone street Lee sees a score of inverted muskets, their bayonets thrust into the ground by soldiers who'd quietly started home during the night.

Lee's commissary officers distribute food from the army's rapidly shrinking supply. Luckier soldiers are given two ears of corn, and that stolen from forage meant for the horses. It's better than nothing, though it makes their jaws ache and their gums bleed. When the supply clerks have no more food to issue, those who've been given nothing wander off to forage for themselves. Many of them won't return.

As the Confederate army prepares to march, its soldiers hear faint, scattered firing off to the northwest. Later, it's whispered through their ranks that Federal cavalry struck a wagon train sent that way ahead of the army. An unconfirmed report but alarming because it would mean that Sheridan's cavalry now are ahead of them.

Lee's soldiers leave the village early in the afternoon, with burning ammunition still popping behind them and a heavy, black cloud hovering over the railroad station. Suddenly Lee hears heavy, rhythmic rifle fire ahead: regiments firing in volleys; Longstreet's regiments.

He hurries Traveller forward to join Longstreet and Lee's son, Rooney.

Rooney Lee: well over six feet tall, heavy, his ruddy face largely hidden by a heavy black beard. Huge hands and feet. Rides one of the largest horses in the army because no other could handle his bulk. Harvard educated where he'd been the most popular man in his class but of whom his professors said, "I never knew him to have an original thought." Not a brilliant man, and appointed a general without ever having led a cavalry charge in the field. Determined to do his best, however, and so far has been reliable in a fight.

"Yankee skirmishers, sir," he reports to his father. "About a half mile ahead of us. They have breastworks about a half mile beyond them, blocking the road. There's cavalry there and infantry. Saw their V Corps' flag awhile back, and Sheridan's headquarters' flag. There's signs of other infantry coming up, but we're not sure which corps or how many."

"Can we get through them?" Lee asks.

Longstreet answers for him. "We can cut through what's ahead of us right now, General. Reckon it's two divisions of cavalry, probably Crook and Custer. Maybe Devin's there too or off to the flanks. And Griffin's corps. If we go now, we can cut our way through, but we don't dare wait."

Lee nods but continues to study the enemy for a long time. At last Longstreet becomes impatient.

I've brought Field (Major General Charles W. Field's division) and Mahone on line, General. Shall I send them in?"

"No. Not yet."

Lee rides a bit farther forward, Rooney Lee by his side; Longstreet from long experience hanging back as Lee works to a decision.

Finally one of Longstreet's staff officers asks, "What's holding him up, General?"

"His blood's up, Colonel. His blood's up. He wants to attack about as much as I've ever seen it. Knows he *has* to attack or try to swing this tired army wide of whatever's ahead of us. And he knows that he just can't afford to be held up here for long. Any more delay and there'll be more of 'em over there. One more Yank corps and we won't have any choices left. Patience, Colonel, patience. He's weighing it all. If he wants us, he'll call us. See, he's turning back now."

"General Longstreet?"

"Yes, sir?"

"It's too risky. I had a report of Federal cavalry northwest of us, near Paineville, earlier this morning. We mustn't delay here any longer. Swing your men northwest," he traces the route on his worn map. "Amelia Springs then Farmville. You will lead. Go through whatever's there. We'll try to get our rations at Farmville then either swing south toward Danville or continue on toward Lynchburg."

"General," he turns to his son, "ask General Rosser to ride ahead of us toward Paineville. Find out what happened there this morning and be certain that the road is clear."

"A heavy march, General," sighs Longstreet.

"Yes; we must do the best we can."

A short while later Rosser confirms that the Confederate wagon train sent in advance of Lee's army has been attacked and burned near Paineville.

He'd found the road, hemmed in by woods and very swampy ground, blocked by the wrecked train. Many burned wagons, with horses and mules dead in their traces and many dead and wounded teamsters.

Led by Major Henry Young's Union scouts, Brigadier General Henry E. Davies' cavalry brigade had slashed at the train's front and rear, scattering its small cavalry escort. Then Union troopers rode among the wagons, shooting down teams and drivers. Several hundred wagons have been burned and six cannon captured.

Lee's infantry can march around the fight, but the dead teams and burned wagons must be pulled to one side before artillery and supply wagons can pass.

CLAIBORNE

Doctor John Claiborne's medical staff are with the ambushed train. Claiborne is riding a strong young mare he'd swapped for his own worn-out horse. Behind him is his buggy, driven by the young slave, Romulus. Behind the buggy, an ambulance-wagon with a wounded Confederate captain and his nurse-sister, a plump Confederate chaplain, and a male nurse named Burkhardt.

All morning Claiborne, hearing firing off to the south, fears an attack on the train. Finally he decides that, while he has time, he'd best do something he's been thinking about.

"Romulus," he calls the boy, "I'm going to give you something very valuable."

Sitting upon a large stone beside the road, he scribbles a note freeing the slave.

"Now, son," he wraps it in a waterproof envelope, "you put this in your shirt pocket there and keep it safe. It's your freedom. No Yankee's going to claim he freed you. I'm setting you free, here and now."

When he's seen the grinning Romulus button the document in his shirt pocket, Claiborne again reaches into his own pocket.

"Here, son; take this money too and this knife to remember me by. Now, you mind your mamma told you to stay with me? Well, you can stay as long as you like but, if the Yankees come and things get mighty hot around here, you skedaddle. Understand? Skedaddle. Don't wait for me; just skedaddle back to your mamma, and tell her that's what I told you to do."

Romulus pockets the treasures and grins again, "Thank you, suh; but I reckons to stay wif you long's I can."

The boy has turned to the buggy and Claiborne is mounting his horse again when firing erupts around them and he hears shouts, "Yankees! Yankees! Everywhere!"

Some teamsters jump from their wagons and begin to run; others try to swing their teams around the stopped wagons. Already the road is blocked. Everywhere men are shouting and horses squealing and rearing.

Perhaps thirty yards away, Claiborne sees the first Yankee raiders, about fifty of them, galloping the length of the wagon train as they fire their carbines and revolvers to down the teams.

He reins his mare into a slight opening in the thick blackjack oak and scrub pine but, when more Confederates follow him, he stops to rally them.

"Turn back, men! There's only a handful of the Bluebellies. We can stop them. Make a stand and save the wagons!"

When the panicky men run by as if they've not seen nor heard him, he waves his arms and shouts again, "Rally here, men! We can stop them!"

Still they run. One soldier, a lean, bearded veteran still carrying his rifle, stops long enough to grin and yell back at Claiborne, "Pilgrim, if you're fool enough to believe that, you go right ahead and stop. I'm headin' for the Tennessee line!"

A Confederate officer joins Claiborne in shouting, "Stop, men! Fire at 'em a time or two and those Yanks'll run away."

He grabs one fleeing soldier by the shoulder, but the man tears away, stammering, "I'd help, Cap'n, but I ain't got no gun."

Claiborne grabs for another, "You man! Why are you running?"

The soldier grins over his shoulder, "Mister, I'm runnin' 'cause I can't fly, and that's the long and short of it. And you'd best run too!"

Climbing on a stump, the captain tries again, "Stop, men! Make a stand! Help me drive those Yanks away."

Then he falls, his arm shattered.

Beyond the woods Claiborne sees a long line of advancing infantry: North Carolinians who drive away Davies' raiders. Too late to save men and teams, wagons, cannon, limbers and caissons. Too late too to save Lee's headquarters' wagon with the records of the Army of Northern Virginia burned and scattered.

Claiborne's hospital train is scattered too. Romulus, Claiborne's surgeons, and the wounded Confederate officer are taken prisoner. The captain's sister, however, has been left with the hospital wagon. She tells him that their former companion, the fat Confederate chaplain who'd ridden the wagon all the way from Petersburg because he said he couldn't walk, led the rush from the fight; like Claiborne's missing terrier, he's disappeared for good.

"That chaplain," she has to laugh, "has to be the fastest runner I ever saw."

Claiborne, his hospital train now reduced to his horse, and a mule-drawn wagon, leads his survivors toward Amelia Springs.

BOYKIN

Major Edward Boykin's 7th South Carolina Cavalry Regiment is among the train's rescuers.

They'd heard the heavy firing. Then they saw the wagon train, nearly two miles long but now with nearly half its wagons burning; caissons and limbers exploding; and a lot of men dead or wounded beside the abandoned wagons.

Most of Davies' Federal cavalry, seeing Confederate riders approaching, flee. A few Yankee stragglers, however, intent on looting the wagons, still remain. Some are wearing Confederate uniforms or other clothing taken from burned wagons. Others have tied silver pots and sugar bowls to their saddles or adorned their horses with clothing, hats, or ribbons they've stolen. So intent upon their work that they don't see the approaching Gray riders.

Then Boykin's men are among them with Colt revolvers and slashing sabers. Most of the looters are shot down or sabered, left beside the road in their captured finery.

Boykin's bugles sound "Recall!" Several miles down the road they've found the rest of Davies' retreating cavalry brigade, and mean to go after them.

Before leaving the gutted wagon train, however, Boykin's Sergeant Major takes him to a beautiful young woman stranded beside the road.

Sobbing but more angry and frustrated than hurt, she explains, "Major, I've been visiting in Richmond but, with the fall of the city imminent, I decided to try to get back to my family beyond Lynchburg. Everything I own is in this wagon. They didn't have time to search it, but they've taken my horse."

She stands with mud to her ankles, a useless buggy whip in her hand, fighting back tears as she begs him to recover her horse.

"Ma'am," Boykin doffs his hat, "we'll do what we can. But now I want you to go with my sergeant to that house up yonder. You'll be safe there until we can come back."

Leading his men down the road, they're rounding a bend when Union soldiers fire from ambush. Several of Boykin's men tumble from their saddles. One screams as his body hits the road, a scream they'll always remember. His back is broken, and he begs them to put him out of his misery. They're lucky; he dies before they must answer.

Another bullet hits Boykin, hard, in the chest; the blow knocking his breath away. He feels himself falling. Then he's able to grab his saddle horn and hang on until his horse stops. Draped over the horse's neck, he fights to remain conscious. Then he feels a hand on his shoulder and someone is pulling him erect.

"You all right, Major?"

"I don't rightly know," Boykin can barely speak, "but I think so. Leastwise I'm not dead yet."

He looks down. A torn, bloody shirt and a deep, sharp pain. Unbuttoning his shirt, he sees blood and an ugly bruise spreading across his ribs. But no bullet hole.

"Whew!" the younger officer exclaims. "You must live right, sir. That ball must've been too spent or you were too tough for it to go through; it ricocheted off your ribs. They'll be sore for a week or so, but you'll never be luckier."

"Here," he passes Boykin his canteen. "Take a pull on this. Picked it up aways back, and it'll help."

Boykin takes a deep swallow. He's not eaten since the night before, and the bullet left him with pain and nausea. So the brandy is doubly potent, coursing its way to his boots. Straightening in his saddle, he returns the canteen with a grin.

"Thanks; that's almost worth getting shot for."

Riding ahead Boykin sees that their Yankee ambushers have joined Union General Davies' rear guard behind a low stone wall.

Seeing the Confederate cavalry preparing to charge, Federal soldiers behind the wall taunt, "Come on, Johnny. Come on. We'll wait for you."

Boykin studies them then commands, "Regimental front, gentlemen." Then he calls out, "Draw sabers!" then "Charge!"

As Boykin's troopers begin their charge, more Confederate cavalry, Rosser's men, suddenly arrive to join the fight.

Union cavalrymen, dismounted behind the wall, fire at the charging riders but have time for only one volley before Confederate riders have jumped the wall and are among them, shooting and sabering, taking few prisoners.

Boykin re-forms his ranks then waves them forward, pursuing Davies' survivors in a running fight that lasts until the Yankee cavalrymen reach the safety of Union General Joshua Chamberlain's infantry lines above Jetersville.

Seeing solid lines of infantry before them, Boykin reins his horse then has his bugler sound "Recall." His troopers have done all they can to avenge the lost wagon train.

Returning to the ambush site, he has the girl's wagon, pulled by a captured horse, driven to the farmhouse. A bad day, but something good come of it.

At sunset Boykin's men camp near Amelia Springs, enjoying some flour and bacon they'd found at a small mill. The first food they've had all day.

Boykin, exhausted and aching from his bullet wound, is rolled up in his saddle blanket when his sergeant shakes his shoulder.

"Major? Sir? Would you like a canteen of old apple brandy? Mighty good stuff, sir. One of the boys found a barrel of it over in the woods. Jack tried it first; you know how he is. Well, after a long pull or two on his own, he let a few of us in on it. We reckon you've earned your share this day."

"Want it? Jim, if we get out of this scrape alive, I'm recommending you for promotion to general!"

The canteen is filled with the same excellent brandy he'd sampled after being hit by the spent ball. Boykin'll never know how the brandy found its way through so many hands, but he feels mighty good that they've shared it with him.

SHERIDAN

Sheridan, anticipating a Confederate all-out dawn attack, is waiting at Jetersville, but the attack doesn't come.

Up the Amelia Court House road they've identified Rooney Lee's cavalry and Longstreet's infantry. Prisoners confirm that the rest of Lee's army is behind them. Still, they don't attack.

As April winds and the morning sun begin to dry the road and fields, Sheridan's and Griffin's men anxiously watch the road, but the attack still doesn't come. No sign of their enemy. No cannon fire; no skirmishers advancing; just an eerie, disquieting silence.

Sheridan paces the center of his line, confers with his commanders and watches for any sign of a forming assault.

And, as often as he looks toward Amelia Court House, he steals glances back over his shoulder, looking for Meade's II or VI Corps. Both delayed somehow, and he needs them right now. Without at least one more Union corps, Lee probably can break his lines then swing south.

No amount of looking or pacing, however, reveals Lee's intentions or Sheridan's reinforcements.

Then Davies' scattered brigade, driven from the burned wagon train, calls for help.

"Griffin," growls Sheridan, "Davies' got himself in a wringer up there. Needs help. Send Chamberlain. Quick now!"

Later, as he watches Davies' relieved troopers pass through the safety of his lines, Joshua Chamberlain shakes his head.

"They've been looting, Ellis. See the silver pots and sugar bowls? And the civilian clothing? Ribbons on their horses? Plunder. Not much discipline there. Not like our Maine regiments, Colonel. Or New York or Pennsylvania either."

Then he growls again, "Griffin says that Sheridan wants to send us back to Meade. That he wants the VI Corps instead. Served with him in the Valley. Well, Ellis, we've served him well too. Don't have to tip our hats to anyone. Be all right with me if he does send us back."

He gets his wish. That afternoon Sheridan will return the V Corps to Meade's command.

Meade arrives at Jetersville to join Sheridan at the Amelia Court House road. He's still so ill that he's ambulance bound. Well enough though to be angry. Angry that, although he ranks Sheridan, Sheridan seems to be in charge.

"I've been waiting for Humphreys (Major General Andrew A. Humphrey's II Corps) and Wright (Major General Horatio G. Wright's VI Corps), General. Thought they'd be here hours ago," a combative Sheridan fires the first barrage.

"Well, Sheridan," Meade steps from the ambulance then steadies himself against its wheel to recover his breath, "if your damned cavalry hadn't cut in front of my men and torn up the roads, we'd have been here a lot earlier. Not to mention the roads we've had to corduroy to get food up here. Food, I might add, for your command too."

Sheridan, taken back by the angry respose from a Meade he's reckoned to be weak, submissive, pulls back.

"Meant no disrespect to you or your men, General. Just anxious."

"Well," Meade scowls, "don't be. Humphreys' right behind me, and Wright's behind him. Where do you want them?"

"I'd recommend you put Humphreys on Griffin's left. That's the greatest danger right now. When Wright's gone in on Griffin's right, I'll pull out my cavalry. Lee's retreating. Swinging north and west of us, and I mean to get ahead of him."

"How do you know that, General? My information has him at Amelia Court House, and I mean to strike him there."

Sheridan describes Davies' morning attack on the Confederate wagon train. Then he pulls a scrap of paper from his pocket.

"Major Young's scouts picked up a courier with this letter from a Colonel Taylor on Lee's staff."

Written that day from Amelia Court House, Meade reads:

Dear Mamma:

Our army is ruined, I fear....General Robert E. Lee is in the field near us. My trust is still in the justice of our cause....General Hill is killed....I send this by a negro I see passing up the railroad. Love to all.

Your devoted son

"Hmph," Meade nods, "sounds mighty discouraged."

Encouraged, Sheridan recovers his briefly subdued spirit.

"I think you should attack. Right now. Up this road to hit whatever's left of Lee's army; meanwhile, I'll send my cavalry west to get ahead of him."

"No," Meade is firm. "We don't have enough strength for that. Perhaps after Humphreys and Wright are fully up and rested. Not before. And as your cavalry's assignment in opening this campaign is done, and there's doubt that we're going to have a fight here, I think that you'd best return Griffin's corps to my command."

Red-faced, Sheridan answers, "As you wish. Anyway, until we see what Grant has in mind for both of us."

Meade, too nauseous to continue the discussion, merely nods.

"I'll set up my headquarters below the village. Meanwhile, as Humphreys and Wright arrive, place them in line for me: Humphreys on the left, Wright on the right, and Griffin in the center. I must rest; we'll talk later."

Later, when they pick up their discussion, Meade still is not convinced that Lee is retreating.

"No, Sheridan," he shakes his head. "I don't think so. His army's still at Amelia Court House. He means to fight there, and I mean to oblige him. In the morning we'll swing east then come in from that direction while you hold here."

"General," Sheridan scowls, "if he's still there, and you swing east, that will put you *behind* him. We must get *ahead* of him."

"If he's there, General Sheridan, it's because he means to *stay* there. I'll order the move to begin early in the morning."

Near sunset, realizing that he can't change Meade's mind, Sheridan decides to appeal to Grant.

He gives his most trusted scout, Sergeant James A. Campbell, a note to be delivered to Grant at Nottoway Station nearly twenty miles away. The note reports Sheridan's operations, including Davies' raid; Sheridan's belief that Lee is moving northwest; and Meade's intent to attack to the northeast in the morning.

His conclusion, the most important part of it all: "I wish you were here yourself. I feel confident of capturing the Army of Northern Virginia if we exert ourselves."

GRANT

Grant has been with Ord's column most of the day. When he learns that Sheridan has blocked Lee's path at Jetersville, he sends officers to spread the news. Men who've already marched fifteen miles that day shout that they're ready for fifteen more.

At dusk Grant hears a commotion ahead. His cavalry escort has spotted a Confederate in the woods ahead and fired on him, the Southerner instantly vanishing.

Then he appears again, now much nearer Grant, waving a once-white handkerchief. A dozen Union troopers instantly converge on him. Then the troop commander reports to Colonel Horace Porter.

"Sir, he says he's one of Sheridan's scouts and that you'll vouch for him. Name's Campbell. Says he has a message for General Grant."

Porter, riding forward, immediately recognizes Campbell: tall, lean, long-haired, wearing the uniform of a Confederate officer.

"Well, how d'ya do, Campbell? How are things with General Sheridan?"

"First rate, Colonel," Campbell spits a long, brown stream of tobacco juice, "leastwise, it 'peared that way when I left him about three hours ago. Got a message for the General, sir. Mighty important."

Campbell, aware that he must cut a pretty good figure in his Confederate uniform, appearing out of nowhere with a special message from Philip Sheridan to U. S. Grant, spits again and sits tall in his saddle. Little enough compensation for the hard, dangerous ride he's had through enemy country, but *something*, particularly when he's reporting before all these headquarters' peacocks. Let 'em see how a real cavalryman does!

It helps too when U. S. Grant himself smiles, offers him a cigar, and shakes his hand. Not bad, Sergeant Campbell, he tells himself; not bad at all.

"You have something from General Sheridan, Campbell?"

"Yes, sir," the lanky scout takes from his mouth a huge chew of tobacco, peels it away, and from its center extracts a small wad of tinfoil. Then he opens the tinfoil and pulls out a sheet of tissue paper: Sheridan's "I wish you were here yourself" message.

Grant reads the message carefully. Sheridan's in a fix with Meade; needs some help.

"Colonel Porter," he's already begun to dismount, "have Cincinnati brought up. He'll be better for this ride than Jeff Davis. You'll accompany me. Pick a couple more officers, reliable ones. And we'll need about a dozen men for escort. Campbell will lead us. It's about twenty miles. Sheridan needs us."

―――――――――

As they begin their ride, Porter's apprehensive about the trip. They'll be riding cross-country, at night; through country they've never traveled before. Country patrolled by Rooney Lee's cavalry. And, despite his greeting Campbell, Porter isn't too sure that the scout can be trusted. He considers most scouts not too trustworthy, and this man has the life of the General-in-Chief of all the Union armies in his hands.

After two hours' ride, down dark lanes and across fields and woodlots without a single orienting feature that Porter can determine, Grant's party completely in Campbell's hands, Porter becomes even more alarmed.

For the last half hour the fence lines they've crossed have had large sections of rails thrown to the ground. Cavalry have passed by, probably not too long ago. And several times Campbell has quietly skirted campfires where they can hear Southern soldiers talking.

When Campbell suddenly drops back then reins his horse into thick woods they'd been skirting, Porter believes his suspicions nearly confirmed. Quietly he reins closer to Grant, drawing and cocking his revolver and holding it across his saddle. If they're ambushed, he means to protect Grant. If Campbell has betrayed them, he also means to kill the scout with his first shot.

Campbell, however, apparently has regained lost bearings and, without looking back, finds and follows another country lane. When they pause again, he turns in his saddle to quietly tell Porter, "Sorry to spook you back there, Colonel. Got myself a mite confused for a minute till I saw that big black oak."

When Porter, feeling guilty for having mistrusted the loyal scout, nods and smiles, Campbell spits again then quietly adds, "You can put up your hog-leg too, Colonel. Won't be needing it here. We passed through our pickets 'bout five minutes ago. See those campfires ahead? They're Sheridan's headquarters. Hope they've saved us some kinda supper."

―――――――――

It's half past ten when they suddenly appear at Sheridan's headquarters, unseen and unchallenged by Sheridan's pickets. Campbell has taken delight in having his whole party suddenly appear as miraculously as he'd appeared before Grant nearly four hours before.

As Campbell expected, cavalry troopers near the little log cabin in the center of a large tobacco field are startled to find their General-in-Chief suddenly among them.

"Why, it's Grant! Boys, this means business!" one exclaims.

Another adds, "Jerusalem! The Old Man's out here himself. Johnny Reb's gonna get busted tomorrow, for certain!"

Or a third, "Grant. Walked right in on us. There'll be lively times 'round here tomorrow."

A shirt-sleeved Sheridan, alerted by the shouts, appears in the lanterned doorway.

"How are you, Grant?" his greeting is as casual as if they were meeting on a busy street in a big city. Not a trace of surprise that Grant is there. Sheridan had said that he wished Grant could come and, knowing the man, he fully expected Grant to do just that. And now he has.

"Sheridan," Grant smiles around his dead cigar. "Reckon you could scare up a bite or two for that man Campbell of yours? He's had a couple hard rides. And I suppose that the rest of us could stand some supper too. Seems you've had yourself a busy day."

A good supper for hungry men: cold chicken, beef, and coffee. Then the generals talk, Sheridan explaining the situation before him and his conversations with Meade.

"We've got 'em, Grant; if we move quickly. If Lee's still at the Court House, and if Ord can get up here early in the morning, we'll be squarely in front of him. If he's already headed west, as I think he is, Meade can pitch into his rear while I take the VI Corps and get ahead of them once and for all. Either way, tomorrow or the next day we can force Lee to fight or surrender."

Sheridan is confident, enthusiastic.

"But," he concludes his report, "Meade won't listen to me, General. He won't listen to me. That's why I asked you to come."

Grant merely nods, puffs his cigar, as he listens carefully to Sheridan's tirade. When Sheridan has finished, he nods once more. Then he scribbles an order for Ord: Watch the roads running south from Burkeville and Farmville; close up on Farmville, and be ready to support Sheridan.

"Lee's in a bad fix," Grant studies Sheridan's map as he muses. "He can't afford to lose time fighting you here. I think, however, that he still can get away by swinging around you. That's what he'll try to do. Probably already started."

"I agree, General," Sheridan glares, again sniffing a fight. "I think he's sent his trains on ahead with the infantry following. It's not too late; we still can bag them all."

"We're doing all right," Grant calms him again. "We've got it all in our favor; but we may not be able to do it in one bite."

He puffs in silence for a moment then stands up. "Come on; let's go wake Meade."

———

They find Meade in his farmhouse headquarters, lying down but, still ill, not asleep. Meade, surprised at finding Grant there, glances suspiciously at Sheridan but says nothing.

When they've discussed the situation before them, Meade reports that his infantry will advance on Amelia Court House at first light. He's modified his earlier plan, however, with his regiments attacking from the west, not the east. Hedging his bet.

Grant smokes a moment, not revealing that he's aware that Meade has changed his plans. Then he nods, "All right. Your orders will hold. But I doubt that we'll find Lee at Amelia Court House. I expect that he's already begun to march around us, probably heading for Farmville. Swing some infantry to the northwest, just in case. The rest can go up this road to the Court House. If they find that Lee has gone, they can turn and catch up. Remember we're trying to head him off, not chase after him. I'll send Sheridan ahead but, until I can get Ord's men closer, you'll have to lend him some infantry."

"Sheridan," he continues, "push hard for Farmville. If Lee's heading that way, you get there first. Meanwhile, his men are going to be awfully tired, strung out. Cut his columns. If you can't stop him from the front, cut him up from the side; slice off some of his units so Meade can finish them off. We'll do this one bite at a time and see what develops."

THE LONG MARCH

At the head of Lee's army, Longstreet is setting the pace for the most difficult march the Confederate army has known.

Lee's soldiers are very tired, very hungry. From the army's last remaining stocks some have been given one day's rations: one-half pound of flour and one-third pound of bacon. Riders, doling out the last of parched corn, meal, or peas to keep their horses alive, sneak a scant mouthful for themselves. Some Gray soldiers have managed to string out the little food they had in their knapsacks, but most of Lee's veterans have no food and no prospect of any until their march ends, somewhere up ahead.

———

Robert E. Lee is invited to eat at "Selma," Richard Anderson's small farm near Amelia Springs. As he's eating, Colonel Marshall arrives with dispatches. Mrs. Anderson, arms on hips, however, won't allow him to interfere with Lee's meal.

"Now, Colonel, that poor man's out on his feet. Needs a decent meal and a few minutes' peace. You leave him be 'til he's finished."

"These are important dispatches, ma'am."

"Not so important that they can't wait just a bit. I'll not have Robert E. Lee leave my home until he's had a meal in peace. Now I've had enough trouble convincing him, Colonel; don't add to my problem."

Marshall pleads, to no avail. Finally, when he sees that Lee is nearly finished eating, he grabs a Confederate flag, wraps it around her waist, and lifts her to one side.

"You see, ma'am; we're on the same side. Now let me see the General."

And she does, satisfied that Lee's had a decent meal and a few minutes' rest in her home; something to tell her grandchildren.

Marshall's messages include a report from John Gordon with the rear of Lee's army. His men have captured two Union couriers with a message from Grant to Federal General Ord, dated "Jetersville, April 5, 1865–10:10 p.m.," directing Ord to watch the roads between Burkeville and Farmville and advising:

I...(believe) that Lee will leave Amelia tonight to go south. He will be pursued if he leaves. Otherwise an advance will be made upon him where he is.

Now Lee knows that Grant is at Jetersville, probably with Sheridan and Meade, and that Ord is at Burkeville. The door is closing on his army; they must march even faster or be forced into a fight they can't win.

Throughout the late afternoon, and evening, and the dark night that follows, the Confederate march is pushed.

The road is strewn with abandoned equipment and personal property. Chalked on a burned-out wagon, some Southern soldier has written, "We'uns have found the last ditch."

Exhausted soldiers sit beside the road. Some will rest then move on to rejoin their regiments; others mean to stay where they are, waiting to be picked up by their Yankee pursuers. Others thrust their bayoneted rifles into the ground then simply head for home.

A lone Confederate straggler, dozing by the roadside, wakes to see a squad of Northern soldiers closing on him.

"Got you, Johnny!"

The Gray veteran looks at his empty rifle, its stock broken; his ragged uniform and worn shoes, both soles parted from their uppers and flapping loose. Then he disgustedly nods and answers.

"Yes, and a hell of a git you got!"

Major Robert Stiles, the praying artillery battalion commander, briefly leaves his men to visit a farm where his mother has been staying.

The farm's owner, a former Richmond businessman, answers his knock.

"Ah, Bob. It's all over isn't it?"

Stiles, still undefeated, shakes his head.

"Over? No, sir, it's not over. We're free of the trenches, free of Richmond. Nothing to be responsible for except ourselves. Now we can get a fair fight in an open field. That's all we ask for. Then let them come on, and the sooner the better."

Their enemy is closer than Stiles realizes. Major Henry Young's Federal scouts are busy delaying the long column of Confederate men and wagons.

Two Gray sergeants wave Confederate Lieutenant Colonel John Haskell's battalion of cannon down a side road where they're ambushed by waiting Federal cavalry. Men and horses are shot down; the survivors, mostly unarmed, scattered. After the raiders have gone, the remaining artillerymen return to find their cannon and mule teams gone, the spokes cut from the wheels of abandoned limbers and caissons. A first sergeant calls his battery's roll, and only ten men answer. Unarmed, they rejoin the main column.

Later it occurs to them that the Confederate sergeants who'd directed them down the road must have been the Union scouts they'd been warned about.

Not far away the 30th North Carolina Infantry Regiment is repeatedly stopped below a ford by a slow-moving, often-stopping wagon train. Finally, their angry Colonel pushes his way to the ford. Two Confederate staff officers are there, insisting that each teamster water his horses at the creek then move forward only far enough for the routine to be repeated by the next teamster.

The frustrated, red-faced infantry colonel confronts them.

"You two are holding up this entire column by watering these teams one at a time. A private of this regiment would know better. Are there no streams ahead where you can water all the horses at once?"

"Reckon not, Colonel. Orders are to do it here."

"Whose orders, sir?" the Colonel bellows. "What damn fool told you to do this?"

"General Pickett, Colonel. And he says to do it here."

"I am of the impression, sir, that General Pickett no longer commands a unit of this army. That he lost his command at Five Forks then skedaddled. At any rate the order is wrong, and I'll tell him so. Where's his headquarters?"

"Up ahead, Colonel. Tell you what, sir; you're on foot. You wait here. We'll ride right up there and check to be sure. We'll be back 'afore you know it."

The two Confederate staff officers disappear up the road, and the slow process continues until the North Carolina colonel decides that the pair, probably Yankee scouts, don't plan to return at all. After that, teamsters aren't allowed to pause at the ford.

Confederate infantrymen lose all sense of time and place. They simply plod along. When the column stops, they stop. Sometimes, if the stopping is prolonged, they sink to the ground for a few minutes' rest. More often, they sleep on their feet. Awake yet not awake; numb to it all. A few yards then stop; then a few yards more.

Sometimes hungry men step from ranks to shoot at nearby animals that have strayed too near the road. Officers who hear the shots look the other way. If a chicken or a pig or a cow is brought down, it's butchered on the spot with men taking what they can of the meat and eating it raw as they march.

It's worse that night. It's cloudy, and the road is dark. Underbrush, close to either side of the road, makes sounds and shadows ominous. And the men are nervous, irritated, confused to begin with.

At a narrow defile a half dozen more of Major Henry Young's Union scouts, hiding in nearby thickets, fire into the Gray column. Panicky Confederates fire back at anything that flashes or moves. Fire into their own men. The uncontrolled fire fight soon involves an entire Confederate regiment. Several dozen men are killed or wounded before the firing can be stopped. Then the dead and hurt men are left by the roadside. Survivors shrug, "Be thankful it's not us; keep moving."

Near midnight, a large, black stallion, tied to a fence rail while its rider sleeps, pulls the rail free. Startled by the rail's dropping but still tied to it, the stallion gallops up the road. The fence rail, flopping behind, crashing into marching men. Unable to see their "enemy" but seeing other men fall, some soldiers blindly fire their rifles. A fire fight erupts for a hundred yards up the column before it can be stopped. Then another dozen dead or wounded men are left beside the road.

Near Amelia Springs, as Lee's army pauses for a short rest, Confederate soldiers hear Federal bands playing in the distance: "Home Sweet Home," "Annie Laurie," "Lorena." Haunting melodies that make it hard for a man to take up the march again.

At dawn, April 6, a footsore, hungry, exhausted soldier from Longstreet's corps shakes an angry fist at the morning sky and sums the night's march:

My shoes are gone; my clothes are almost gone. I'm tired; I'm sick; I'm hungry. My family...killed or scattered. I've suffered all this for my country. I love my country; I would die—yes, I would die...(for) my country. But if this war is over, I'll be damned if I'll ever love another country!

LINCOLN

In Richmond, the *Malvern*, with Abraham Lincoln aboard, is anchored at mid-stream off Rocketts Landing.

Admiral David Porter's fears of an assassination attempt are renewed this morning by a report from Colonel Edward H. Ripley, Weitzel's commander in Richmond. Ripley brings to the *Malvern* a Southerner named Snyder who claims to have served with Rain's Torpedo Service, a Confederate secret organization responsible for a huge explosion at City Point months before.

Now Snyder claims that he knows of a planned secret mission "aimed at the heart of the Yankee government." To Snyder this means an assassination attempt on the President; he's come to warn him.

When Porter and Ripley tell Lincoln, the President nods his understanding but declines to see Snyder.

"You see," he explains, "I've received so many of these reports that I no longer know which to believe and which to discount. So I just file them in an envelope I've marked 'Assassination.' It may be true, gentlemen, and I appreciate your efforts to protect me, but I must go on as I've begun. I don't think that anyone would intentionally harm me, but the course is laid out and I must run it. Still, it would be foolish not to take the precautions you feel best."

While Porter is ashore investigating Snyder's story, three more visitors come to the *Malvern*: Judge John A. Campbell; Union General Godfrey Weitzel; and a prominent Richmond attorney and former member of the Virginia legislature, Gustavus Myers.

Lincoln spends several hours with them and, when they leave the *Malvern*, he shares their discussion with Porter.

"Admiral, I've given Judge Campbell a letter repeating the three conditions necessary to end this war: restoration of the Union; an end to slavery; and the disbanding of all Confederate armed forces. I couldn't agree to the general amnesty for Confederate leaders they wanted, but I promised to use my pardoning power to save any repentant sinner."

Then he concludes, "I also gave them permission to convene the Virginia legislature here as soon as possible. Judge Campbell and Mr. Myers promise they'll have Virginia voted back into the Union and that all Confederate troops under General Lee will be ordered to lay down their arms. If they do that, particularly if General Lee is part of the action, the other Southern states will follow. We'll end this war, Porter. And we'll do it without the last horrible battle General Grant fears we must endure."

"That's fine, Mr. President," Porter reflects, "except, if I may point out, sir, you've given the Rebel legislature permission to assemble and conduct business while Richmond's under martial law. If you allow any civil authority to function without General Grant's approval, you'll undermine his position. It could cause embarrassment."

"But," Lincoln is astonished, "General Weitzel was there and did not object, and he commands here."

Porter shrugs, "Perhaps he missed the fine point, sir. Anyway, you're our Commander-in-Chief and can do whatever you feel best. But you've been very careful not to conflict with Grant's authority. Perhaps you should suggest that they request Grant's permission to meet."

"It's more complicated than I thought, Admiral. Thank you for catching my error. Please go recover the letter I gave Judge Campbell. Tell him I'll provide him with another."

When Porter returns with the letter, Lincoln amends it, advising the legislature to seek Grant's permission to meet.

He sends a copy of his revised letter to Grant, along with his personal note: "Nothing that I have done or probably shall do, is to delay, hinder, or interfere with you in your work."

That afternoon he has another visitor. A large, burly, rough-looking man in homespun wool and carrying a long, heavy walking stick appears on the Rocketts Landing wharf to bellow.

"I'm Duff Green, and I want to see Abraham Lincoln. My business concerns me alone; you just tell him that Duff Green wants to see him."

When Porter relays the message to Lincoln, the President sighs, "All right. Bring him on board. He's an old friend, Admiral; went south with the rest of the Confederacy but still a friend. I'll talk with him."

Porter is waiting on deck when Duff Green clambers up the *Malvern*'s ladder, scowling and angrily repeating, "I want to see Abraham Lincoln."

"When you come in a respectful manner, Mr. Green; and when you've thrown away that cord of wood you're carrying, I'll take you to the President. Not before."

"Hmph," snorts Green, scowling again at the Federal flag on the *Malvern*'s stern, "And how long has Abraham Lincoln taken to aping royalty? I expect haughtiness from a naval officer, but not from a man of Abraham Lincoln's horse sense. Or has he lost that too since last I saw him?"

"Throw the staff overboard, Mr. Green, and mind your manners or I'll have you thrown after it."

Green curses, growls, then throws his staff into the James.

"Is he afraid of assassination? Tyrants generally are."

"You wait here, Mr. Green. The marine guards will see that you don't misbehave and, if the President still wishes to see you, I'll take you to him."

When Porter reports Green's odd behavior Lincoln smiles sadly and nods, "Yes, I fear that he always was a little addle-pated. I shan't mind him. Let him come down."

"I must insist on being present, sir."

"Of course, Admiral," Lincoln smiles, "and thank you for your concern."

When Porter returns with Green, Lincoln rises with an outstretched hand. Duff Green, however, ignores it, immediately launching into a tirade.

"No, I'll not shake your hand, Lincoln. Perhaps once, not now. I won't touch it. It's red with blood. It has cut the throats of thousands of my people; and their blood, now soaking in the ground, cries to heaven for vengeance."

"I didn't come here for old times' sake," he continues, "but to speak my mind. You've come here, protected by your army and navy, to gloat over the ruins and desolation you've caused. You are a second Nero, Abraham Lincoln. If you'd lived in his day, you too would have fiddled while Rome burned."

Despite the storm of abuse, Lincoln still smiles and once more offers his hand. The smile, however, fades as Green continues to ignore the outstretched hand and to abuse him more. Obviously the man has nothing constructive to offer but is only venting his temper.

When Green's voice has risen to a shout, his huge body shaking in anger and his fist knotted below Lincoln's face, Lincoln finally has had enough.

"Stop it! Stop this nonsense, you political tramp!" he orders, now gesturing with the hand he'd offered in friendship. "You sided with those who've tried to ruin our country, but you never had the courage to risk your own precious hide for the things you wish on others. You're just here to pick up any crumbs that may fall your way. You had no principles in the North, Duff Green, and you've brought none to the South."

His voice, not as loud as Green's but firmer, Lincoln continues, "You've all cut your own throats, sir. Unfortunately, you've also cut the throats of many more of your poor followers. Go, you miserable imposter; go before I forget myself and the high office I hold. Go! I don't want you desecrating this national vessel another moment."

When Green stalks from the cabin, Porter follows him to instruct the Officer of the Deck.

"Take this man ashore and see that he does not again set foot on this or any other naval vessel."

When he returns to Lincoln's cabin, Porter finds the President calm, but his countenance is heavy and depressed.

"Mr. President," Porter suggests, "why don't we go back to City Point? You've done what you can here. We can go ashore there among friends, and you can read Grant's latest dispatches and be with those who understand the significance of this week."

"Yes, Admiral," Lincoln agrees. Then he smiles, "I do seem to be putting my foot in one bear trap after another. If Seward were here," he chuckles aloud, "how he'd lecture me for giving Campbell permission to reconstitute their legislature. I'd never hear the end of it."

Returning to City Point they pass a Federal ship bearing more than a thousand Confederate prisoners.

The prisoners, somehow knowing that Lincoln is near, rush to its railing for a look at the President.

"They seem all right," Lincoln returns their waves. "At least they have some meat and bread. I expect that's more than their companions still in the field have this evening."

The Confederate prisoners, seeing Lincoln's smile and wave, continue to grin back and to wave grimy caps and handkerchiefs.

"That's Old Abe," one calls. "Give the old fellow three cheers, boys."

Another calls, "Hallo, Abe. Your meat and bread's better'n the popcorn we're 'customed to."

Lincoln gazes at the outgoing transport for a long time then reflects, "They'll never shoulder a musket in anger again, Porter. I wish that Grant could end it for the others that way. But, if he's wise, he'll let them keep their guns to shoot crows with and their horses to plow with. It would do no harm."

CHAPTER 4

Sayler's Creek   **April 6, 1865**

THE RETREAT

Through the rainy, cold spring morning Lee's soldiers struggle toward Rice's Station on the Southside Railroad. Southwest from Amelia Springs, where they'd briefly rested the night before, then, five miles northwest of Rice, Farmville where their desperately needed rations may be waiting.

The drawn-out Confederate army is more vulnerable than ever. Somewhere nearby are Sheridan's cavalry, a constant threat to block their retreat or to slash at Lee's long, thin line of wagons, cannon, and infantry. And, closing from the south and east are Meade's Army of the Potomac and Ord's Army of the James. Should either catch up, that could end it all.

Lee's line of march is marked by broken wagons and ambulances; by guns and caissons abandoned because the horses no longer can pull them or because untended wheels have finally given out; by discarded muskets, blanket rolls, canteens; by stragglers hunkered down by stone walls or split-rail fences beside the road, making no attempt to hide from officers and sergeants who look the other way anyway; and by Gray soldiers stumbling through woods or across fields: some searching for food, others simply heading home.

Still, within Lee's ranks there are determined veterans whose rifles are loaded and whose bullet pouches are filled. Veterans who encourage the weaker-hearted that maybe just over that next long hill they'll see Rice's Station and maybe even the food they've been told will be waiting. A little food and a couple hours' rest, they say, and it will be all right.

Some Confederate officers seem dazed, confused, exhausted. Robert E. Lee, however, is as vigorous and alert as ever, ordering John Gordon to continue protecting their rear and to burn all the bridges behind them.

All day Gordon's corps will skirmish with Humphreys' pursuing Federal II Corps; Gordon's infantry crouching behind stone walls to fire on advancing Union skirmishers, forcing Humphreys to deploy his regiments. Then, having delayed their enemy, Gordon's men pull back and allow the next brigade to take up the fight.

If a man is wounded, he's given a full canteen of water and a little food, if any is available, and left beside the road. Confederate ambulances can carry no more, and the pursuing Yanks will pick up the Confederate wounded; it's the best Lee's officers can do.

THE PURSUIT

Early in the morning Meade's infantry cautiously move northeast toward Amelia Court House where Meade anticipates finding Lee's entire army.

About 8:30 a.m., however, an officer of the 26th Michigan Infantry Regiment catches a quick glimpse of the last of John Gordon's rear guard disappearing up the Amelia Springs Road. He alerts his colonel and, a little later, Meade, finally convinced that Lee does not plan to fight at Amelia Court House, swings his Army of the Potomac in pursuit.

HIGH BRIDGE

Soldiers of Major General Edward O. C. Ord's Army of the James are marching west, below Meade, to support Sheridan's cavalry if they get into the fight the little general's trying hard to bring on.

At Burkeville, Grant gives Ord a second, special assignment.

"General, you see these bridges across the Appomattox?" he points on his worn campaign map. "Well, I want you to get to them, particularly this one, 'High Bridge' the locals call it; and the Farmville bridge. Get to them ahead of Lee and burn them. See if we can't stop Lee at the river or at least slow him down so Meade can come up on his rear."

High Bridge, the first and most important of Ord's objectives, is three miles northeast of Farmville. It's more than 2,500 feet long; and, where it spans the Appomattox, 125 feet high. A massive bridge of heavy wooden trusses resting on twenty-one tall, brick piers; considered the largest bridge in the world. High Bridge: the most critical bridge on Lee's line of retreat.

Ord forms a raiding party of nearly 900 men: the 123d Ohio Infantry Regiment, the 54th Pennsylvania Infantry Regiment, and three companies of the 4th Massachusetts Cavalry Regiment. Its overall commander will be Lieutenant Colonel Horace Kellogg, with its small cavalry detachment led by Colonel Francis Washburn.

Kellogg sets out immediately, passing through Rice without seeing a sign of their enemy. Then he swings north toward High Bridge.

Southern farmers have spotted the raiding party and alert Confederate General James Longstreet, northeast of Rice at the head of Lee's hardmarching army. Recognizing the danger, Longstreet immediately turns to General Thomas L. Rosser.

"Rosser, round up every man and every horse you've got. Go after them and, if it takes every man you have, stop them. Don't let them destroy that bridge. Go; go quickly. You must save that bridge!"

The Union raiders' march is slowed by their infantry, and several miles below the bridge they're overtaken by Brigadier General Theodore Read from Ord's headquarters who warns them that Confederate cavalry have been seen nearby.

Read and Kellogg confer. Despite the warning, they've seen no sign of their enemy. And the bridge is an easy ride away; a plum ripe for picking. Read makes the decision: Kellogg, with his two regiments of infantry, will block the road below the bridge. If enemy cavalry come, they're most likely to come from that direction. Meanwhile, Colonel Washburn's eighty-man cavalry force will hurry up the road, burn the bridge, then return as quickly as possible. With luck, they can get the job done and be out of there before enemy cavalry can reach them.

Around noon Washburn's cavalrymen reach the Sandy River, their last natural obstacle before High Bridge. On a hill beyond Sandy River, Washburn sees Confederate breastworks, including four cannon.

While Washburn keeps their enemy, homeguardsmen from Farmville, busy, Lieutenant George F. Davis of the 4th Massachusetts Cavalry Regiment fords the little river to strike the Confederate flank. A half-hour's fight then their enemy withdraw. The road to High Bridge is open. Washburn, however, has lost valuable time.

Before he can attack the bridge, he hears heavy firing behind him: Confederate cavalry attacking Kellogg's infantry. He turns to go to their aid.

After quickly conferring with Read, Washburn decides to charge their enemy with his own 4th Massachusetts' cavalry.

"By Fours to the Left!" "Gallop!" "Charge!" his voice rings across an open field, and his eighty horsemen swing to strike the flank of Rosser's dismounted cavalry.

They've no sooner struck Rosser's line when Confederate Generals Munford's and Gary's cavalry brigades thunder in to join the fight.

Union and Confederate cavalry, the Massachusetts' men greatly outnumbered, collide at full gallop. Men are shot or sabered from their saddles. Horses scream and tumble. Solid lines of charging men dissolve into many small, deadly brawls: fierce hand-to-hand fights with carbines, sabers, and revolvers at very close range.

Confederate Brigadier General James Dearing encounters Washburn. Their sabers clash loudly, again and again, as each tries desperately to slash the other from his saddle. Then, in the melee, the two are separated by other fighting cavalrymen.

Washburn, wheeling his horse about, is shot in the mouth and falls in the dusty road. He tries to rise but then is sabered by another Confederate horseman. Mortally wounded, he lies in the sandy trace while the fight rages around him.

Dearing fires his revolver at Union General Read, and Read falls, dead, on the road. At the same moment, Dearing is shot by another Northern soldier. Washburn and Dearing, who'd been cadets together at West Point, lie almost side by side in the hoof-torn road.

A Union saber slashes Rosser's arm, but the Confederate leader manages to stay in his saddle.

Major John Locher Knott, commanding the 12th Virginia Cavalry Regiment, is killed; within thirty yards of his body lie two other Confederate regimental commanders, also killed in the brief fight.

More Confederate cavalry pile into the fight, scattering the 4th Massachusetts Cavalry and breaking Kellogg's infantry line.

Sensing victory, Confederate soldiers shout their Rebel yell and fight harder, driving their demoralized enemy back toward High Bridge and the bluffs above the Appomattox. There, with a steep bank behind them and nearly surrounded by their screaming enemy, the last of the Union raiders surrender.

Rosser has captured close to 800 Northern soldiers, six Union flags, an ambulance, and a large brass band that somehow had found its way into the raiders' party.

Besides the hundreds taken prisoner Federal losses include General Read dead and Colonel Washburn mortally wounded; the 4th Massachusetts with 15 men killed or wounded; the 123d Ohio with six men killed or wounded; and the 54th Pennsylvania with 21 men killed or wounded.

Confederate losses include General Dearing and about a hundred more officers and men killed or wounded.

Dearing, carried to a house near the field and barely able to speak, asks Rosser to take the stars from his collar and give them to Colonel E. V. White of the 35th Virginia Cavalry Battalion as a reward for White's courage during the fight.

Nursing his wounded arm and riding a different horse and wearing a different saber than he'd had early that morning, Rosser reports to Longstreet that he's killed or captured every Union raider except a few who'd escaped by swimming the Appomattox. High Bridge has been saved for Lee's army.

"It was a gallant fight, General!" Rosser exalts. "This is Read's horse; and this is his saber! Both beauties, aren't they?"

Longstreet, to whom such trophies long ago had lost their appeal, wearily nods.

"You did well, Rosser. A complete victory. But dearly bought, General, dearly bought. Dearing, three regimental commanders, and nearly a hundred other men killed or wounded."

"Well," he sighs, "see to your men and horses and picket the town. General Lee's sent word to Farmville for them to fetch us a train of rations, but we must hurry. He was here but heard that Gordon's in a fix five or six miles above us and no word at all from Anderson or Ewell. He's back there with Mahone now, tryin' to figure it out."

SAYLER'S CREEK

Longstreet's corps, including the remainder of A. P. Hill's old Third Corps, lead the Confederate march. Anderson's corps, with the remnants of Pickett's division, follow. Then Ewell's corps: Kershaw's division and the Richmond Reserves. Behind Ewell: a long line of wagons; then John Gordon's Second Corps protecting the army's rear.

Their line of march: Amelia Springs to Deatonsville to Rice's Station, about twenty-three miles.

About five miles above Rice they turn at Hott's Corner to march west along a sandy road that passes the Hillsman farm on the north slope of the valley of Sayler's Creek. The road then descends to cross the creek before doubling back to climb the valley's south slope then swing south across rugged, hilly, wooded ground toward Rice.

Longstreet crosses Sayler's Creek without trouble. The long column behind him, however, slowed by the poor road and by Union cavalry's hit-and-run stabs at Anderson's and Ewell's wagon trains, have allowed a dangerously long gap to develop between Longstreet and the rest of Lee's army.

About 11 a.m. Crook's and Custer's cavalry strike Anderson's and Ewell's wagons near Hott's Corner, the little crossroads a mile east of the Hillsman farm. It's more an annoyance than a serious threat but, while Anderson pauses to beat them off, the gap between his command and Longstreet's widens even more.

When they've driven off Sheridan's cavalry Anderson's men resume their march, following Longstreet's trail past the Hillsman farm, across Sayler's Creek, then up the valley's south slope.

Anderson is near its crest when his scouts call him forward.

"Yank cavalry, General. A lot of 'em. Crook's flag's up ahead of us and on our left flank over yonder; and Custer's got himself a lot of men in the woods over to our left. Can't miss the man's long hair. Don't see any Yank infantry, just cavalry. Hear their band playin' over in those woods? It 'pears the Bluebellies don't mind our knowin' they're there! Looks like they're fixin' to attack, and there's more and more of 'em all the time. Reckon they'll come on when they're ready. No sign of Mahone's men in front of us. Reckon he's up ahead somewhere with Longstreet's people, but we don't know how far or what they've run into. Figured you oughta know 'bout it."

Anderson nods, waves his arm to halt the column, and sends a messenger to bring Ewell, the senior Confederate officer on the field, forward.

"Cavalry, General," he reports. "Lots of it. Crook's for certain; and Custer's over there in strength with Devin coming up. No infantry in sight, but they're probably hurrying this way. Mahone, somewhere ahead of us, doesn't know that we've had to stop, and he's gotten himself out of touch. Seems we're on our own."

Ewell, looking very uncertain, nods, holding his head in that odd turned-askew way that reminds others of a bird turning its head to hear. Shifting his weight on his gray horse, Rifle, he considers the situation. Forced by his artificial leg to be strapped to his saddle, Ewell is hurting. And he's very tired.

"Well, what do you recommend, General?"

"We have two choices, sir," Anderson answers. "We can cut our way through whatever's ahead of us; or we must find another road that swings right toward Farmville. I recommend that we attack, as quickly as possible, and clear the road ahead of us."

"Well," Ewell answers, "I'd be inclined to go around them, through the woods; but you know the ground, and I don't."

Anderson, knowing the ground only by his being on the dirt road a mile ahead of Ewell before being stopped by his scouts, says nothing but waits for Ewell's decision.

Ewell's answer is puzzling. As if his mind has wandered from the critical situation they face, he remarks, "Tomatoes are good. I wish I had some."

Then they hear artillery fire across the valley: the end of Ewell's column or perhaps John Gordon's men have been engaged.

"Do as you think best, General," Ewell reins his horse back down the road. "I must see what has happened behind us."

As Ewell can't decide what to do, Anderson turns to his commanders.

"We'll fight, gentlemen. General Johnson (Major General Bushrod R. Johnson), put your division on the right; General Pickett (Major General George E. Pickett, now without a command), take the rest of our troops on our left. Set our line along the edge of the woods there, where we'll have clear fields of fire. We'll have to watch our right flank very carefully. I'll put some scouts there. Not much in the way of reserves; you'll have to stop them with what you have. Then we'll see what develops."

As Anderson prepares to fight where he is, Ewell, hurrying back down the road, hears the deep bark of cannon fire and ragged, tearing rifle volleys before him.

It's not Gordon's men. Gordon's no longer behind him. Following the fight at Hott's Corner, as Anderson turned west toward the Hillsman farm, Ewell ordered the long wagon train behind Anderson to swing north to a second, parallel dirt road about a mile above Hott's Corner. That, he reasoned, would give the slow, cumbersome, vulnerable wagons a better chance of avoiding Sheridan's attacks from the south and allow Gordon's following infantry to close up on Ewell's column in the valley.

Somehow, however, he'd forgotten to tell Anderson and Gordon of the change. When Gordon reaches Hott's Corner and sees that the wagon train he's to follow has continued north, he shrugs his shoulders and follows it.

When Humphreys' pursuing Union II Corps' skirmishers reach Hott's Corner, they too march north, after Gordon's infantrymen.

Behind them, when Wright's VI Corps' scouts reach the crossroads, they halt, study the ground.

"'Pears a lot of infantry and wagons went straight ahead, Colonel, whilst the rest of 'em turned west. The II Corps must've gone north after the first bunch. Reckon we oughta pitch into this bunch west of us?"

"Yes," the colonel decides, "and let's move faster until we make contact." Wright's skirmishers soon catch up with Ewell's rear guard, and a fire fight develops near the Hillsman farm.

Most of Ewell's men and wagons have crossed Sayler's Creek and are well up the valley's south slope when they hear the first burst of rifle fire behind them. Then the firing, the bugle calls, and the shouts of fighting men grow louder, shattering the afternoon calm. Across the valley, at the Hillsman farm, Union infantrymen are charging down the road, trying to break through the Confederate rear guard to get at the lumbering Confederate wagons at the bridge.

A Union artillery battery wheels into position at the Hillsman farm then joins the fight, lobbing shells down the road toward the Confederate wagon train struggling to cross the creek by a single, flimsy bridge. Hot iron fragments rain down through young spring leaves to strike men and horses as other Federal batteries join the fight.

Anderson, Sheridan's cavalry before him and threatening his left flank, still far from Rice, is spreading his regiments to fight. A mile behind him, Ewell, his rear already under attack, is forming a separate battle line above the valley's floor. And, in a third action another mile to the north, John Gordon already is fighting hard to hold back Humphreys' Federal II Corps while he struggles to get more wagons and cannon across two flimsy bridges spanning the upper branches of the creek.

At the crest of the valley's south slope, Anderson's officers and sergeants carry out their general's decision to fight where he is.

"Hurry up, hurry up!" they call. "Tear down that fence; throw up breastworks along this wood line. Pile on whatever you can find. Put everyone who has a rifle and ammunition in the front line. Have the rest of the men lie down behind them. When you see a man fall, have one of 'em take his weapon and use it. Lots of Yanks out there; lots of Yanks for all of us! You men: take blankets back to the wagons, load up with cartridges and caps, and hustle back here. Doctor, set your station back there in the woods. Quick now!"

Anderson places General Bushrod Johnson's men on his right; General George Pickett's on his left. Johnson, he figures, will be all right. Pickett, on the other hand, is a questionable asset. Pickett: unable to shake off Gettysburg where he'd taken nearly 6,000 casualties in forty-five minutes; then Five Forks, four days ago, where he'd lost the right flank of Lee's army and brought on this retreat. Pickett: present but leaving most of the decisions to his subordinate commanders. No reserves; if the Confederate line breaks or is flanked, Anderson's fight will end quickly.

As Anderson's men brace themselves to meet the Federal attack, several hundred yards down the slope before them Crook's, Devin's, and Custer's cavalry are forming their lines. Two brigades of Crook's cavalrymen have dismounted to charge on foot. The rest of the cavalry will be on their flanks.

At regular intervals in the Federal lines there are splashes of bright colors: red, white, and blue National and regimental flags; and scarlet, white, and yellow cavalry-forked guidons. A lot of them.

Another time Blue and Gray soldiers might have enjoyed the otherwise peaceful, beautiful scene before them: a long, open slope, stitched across its shoulders by a weather-faded gray split-rail fence, descending into a pleasant valley. In the valley, thickets and a small, meandering stream, then a similar slope climbing toward the Hillsman farm.

A valley until today untouched by the war: never disturbed by the shouts of men, the screams of hurt horses, the chilling whistle of shot and shell. All that, however, quickly changing.

Another alien presence, several of Custer's regimental bands, on their all-white or all-gray horses, are cheerfully tooting away: "Yankee Doodle," "Garry Owen," "When Johnny Comes Marching Home", stiffening Yankee cavalry to charge across that field. Forever changing the scene.

"It's a long ways up to those woods, Bob," a nervous young Pennsylvanian about to make his first cavalry charge shifts in his saddle and, as if their enemy several hundred yards away can hear him, speaks quietly to his messmate.

"Sarge says that's where they're at, along the edge of those woods. Can't see 'em 'cause of the slope and that fence up there. Reckon they can't see us either?"

For the third time in as many minutes the youth adjusts the bright red neckerchief already tight around his throat. It's like the one Custer wears, and the whole brigade recently adopted it for their own.

Then the boy confides, "Didn't plan on seeing the elephant chargin' across an open field this way, Bob. Uphill too. Don't know what I expected, but this ain't it. Look, there's Crook on his mule. Looks like a seedy, old preacher, don't he? Calm as if it's a Sunday walk. Swears by mules, they say; says they're smarter'n horses. Reckon it's so? And look, here comes Custer again. Been back and forth a half dozen times already. Right out in the open, like he's not afraid of nothin'. All excited too, like he can't wait for it to start."

"Well, you two quit frettin' over it," his sergeant interrupts the youths' nervous chatter, spitting a long brown stream of tobacco juice at a nearby tree.

"And stop that jawin'. Thinkin' 'bout it just wears you out. Could be worse. Devin's over on our left, and the Cap'n says Crook's sendin' a couple brigades in on foot. That'll keep the Johnnies busy. While they're lookin' to Devin's and Crook's boys, we'll hit 'em right here. Now, once we start up this field, you two keep on goin'. No matter what. They've got one shot in those Enfields of theirs; you've got seven in the Spencers you're carrying. You don't get hit with that one shot, you'll be into 'em before they can reload. Nothin' to it. Now check your weapons again, and watch for my signal. Careful jumpin' that fence up there; your horses'll be gettin' tired. Stay close to me."

Somehow, before the youths feel quite ready for it, a single bugle sounds. Then a dozen more. All across Custer's line. More bugles. Then the sharp, staccato notes of "Forward!" "Gallop!" "Charge!" sound across the long, open field. Nervous, impatient horses, recognizing the calls, need no urging. Whether their riders want to go or not, they start up the slope.

Then the first Confederate artillery fire strikes them. Solid shot plowing down the slope, tearing into men and horses; explosive shells bursting above them and in their faces. Sharp iron shards flying everywhere. Men and horses struck down. They've not yet reached the split rail fence, and that still a long way from the Confederate line.

Crook's dismounted cavalrymen reach the fence straddling the slope.

"Too strong to pull down, boys; go over it," someone calls, and they follow the voice. As they scramble over its rough rails, they're hit by solid sheets of rifle fire from Anderson's lines.

From the woodline above them one Confederate rank fires while the next loads. A solid line of rifles aimed at the knees of the blue uniforms crowded before the fence. Blue-uniformed lines that vanish in the powder's thick smoke. Most of the charging Federal line is down; their second line staggered. Then they're hit by a second Confederate volley, as bad as the first.

Crook's attack stutters as his dismounted troopers stop to angrily return the hail of fire. Unable to see their enemies, they fire blindly into the smoke before them.

Confederate soldiers fire another volley, and another. Crook's attack is broken, and his cavalrymen turned infantrymen fall back.

A few of them cling to the fence line until Confederate sharpshooters, concentrating on these more stubborn Yanks, shoot them down.

Anderson's riflemen have stopped the first Federal charges, and both sides count their losses; regroup; prepare for the next assault.

Urged by shouting sergeants, pairs of soldiers hurry shelter halves or blankets heaped with cartridges and caps along their respective lines. Men fill their bullet pouches and their pockets; raise canteens to ease parched throats, splash a little of the precious water on faces grimy from burnt powder; and get ready to fight again.

Crook, Devin, and Custer are all repulsed. But they've felt out Anderson's line and don't mean to be repulsed again.

This time Crook shifts Brigadier General J. Irvin Gregg's brigade to strike Anderson's right flank. When they've turned that flank, Custer's cavalry will charge the Confederate front.

Northern bugles sound again, and nearly 1,500 Union cavalrymen hit Anderson's far right, hard, and there aren't enough Confederate soldiers there to stop them. The thin Gray line is thrown back into Anderson's wagon train. Gregg's troopers strike the train from the right, shooting horses and drivers as they sweep behind the Confederate line.

––––––––––––––

Behind their hasty breastworks at Anderson's center, a thin line of riflemen watch as Custer's men again charge up the slope.

When the galloping cavalrymen are almost on top of them, the defenders, shouting their Rebel yell, fire a pointblank volley. Michigan and Pennsylvania cavalrymen are shot from their saddles, horses tumbled. Not enough fall, however, to break the charge and, before Anderson's men can reload, Custer's troopers are among them, firing their revolvers and slashing with their sabers.

All with cannon roaring, shells screeching, men cursing and shouting, and Custer's massed bands, slowly following the charging cavalrymen as they play "Garry Owen", Custer's personal favorite, for all their worth.

––––––––––––––

Troopers of the 1st West Virginia Cavalry Regiment, many of them riding mules they'd recently swapped for their own worn horses, are among the first Union cavalrymen to reach the Confederate line.

Some of Custer's men have given the 1st West Virginia a hard time about their mules, seeing the long-eared, cantankerous steeds as a poor substitute for the horse. As if the maligned mules understand this, they lead Custer's charge, easily clearing the split-rail fence then, ears laid back, scampering over the Confederate breastworks.

Sergeant Francis M. Cunningham's mule lands beside the wide-eyed color bearer of the 12th Virginia Infantry Regiment, and the two men struggle for the Confederate flag until the Virginian falls.

––––––––––––––

Captain Tom Custer, only a few yards ahead of his famous general-brother, also is among the first to leap the Confederate barricade.

With heavy firing all around him, Tom Custer reins his horse to face a Confederate soldier carrying the flag of the 2nd Virginia Infantry Reserve Battalion.

"Surrender, Johnny! Surrender and give me that flag!" he shouts.

Instead the Southern soldier snarls, "Damned if I will, Yank," and fires his pistol, pointblank, at his attacker's face.

The bullet tears through Tom Custer's left cheek to break out the back of his neck. His face bleeding and peppered with burnt powder, he reels in his saddle but recovers to shoot down his enemy. Then he seizes the Confederate flag.

Behind him General George A. Custer leaps the same barricade. As Custer's horse lands, it's tumbled by Confederate fire. Custer staggers to his feet to find his younger brother anxiously looking down at him.

Tom Custer, with blood gushing from his blackened face, gasps, "Aut! You all right?" Then, "The damn Rebs shot me, Aut; but I got their flag!"

Then, still waving his captured flag, he whirls to pursue Confederate soldiers scattering into the woods. Before he can gallop off, however, his brother grabs the horse's reins.

"Wait a minute," George Custer shouts. "Where do you think you're going?"

"After Ewell. We've got 'em on the run, Aut. We can bag all of them."

"Tom, you crazy fool," Custer half angry, half amused, laughs, "you're bleeding like a stuck pig. Get back across that field and get yourself patched up. We can handle this without your bleeding to death. Go find a surgeon."

"Is that an order?"

"You bet it is; and you damned better obey it, Captain."

Grinning, he turns to an aide, "Lieutenant Christiancy, take Captain Custer, in arrest, to the dressing station. If he puts up an argument, shoot him again!"

For his action at Sayler's Creek Tom Custer will be awarded the Congressional Medal of Honor, the second time he's won that award within a week.

Custer's cavalrymen, having broken Anderson's line, sweep across the rear of Anderson's and Ewell's embattled corps.

A Virginian, William L. Timberlake, fires his last shot and throws down a rifle so hot it burns his hands.

A Yankee sergeant waves him to the road where other prisoners are being gathered. There's a ditch there and, terribly thirsty, Timberlake uses his hands to lap its muddy water. Laps it as fast as he can until, his thirst eased, he tastes the water's sweetness. Blood. And nearby the bodies of a half dozen Northern soldiers tumbled into the ditch as they'd crossed the road.

He's thinking of that and of the hopelessness of their fight, when one of Custer's cavalrymen, a graying man who wears sergeant's stripes on his sleeves, reins his horse and, seeing the Virginian's haggard face, quietly asks, "You hungry, Johnny?"

"That's a good question to ask any Rebel, Yank!"

"Never mind, boy. You got a knife?"

"Yes."

The Union sergeant turns his horse. Strapped to its saddle is a large ham.

"Here, cut yourself a piece, Johnny. Take a big one. And there's some hardtack in this saddlebag. Looks like you could use a drink of my canteen too."

When his former enemy has turned away, the Virginian, tears streaming down his grimy cheeks, calls after the rider, "God bless you, Yank. God bless you!"

The sergeant waves an arm in farewell, continues down the hill. Timberlake will never see the man again nor know his name, but he won't forget him.

After the Sayler's Creek fight, Custer will report capturing about three hundred wagons, fifteen cannon and several thousand prisoners.

EWELL

Meanwhile, Ewell has gotten most of his wagons across Sayler's Creek and ordered his men to throw up breastworks across the road.

Kershaw's veterans are on his right; Custis Lee's Naval Brigade and the Richmond Reserves on his left.

Kershaw's men will fight hard; always have and so have a reputation to maintain. Ewell's not so sure, however, of the Richmond Reserves: artillery gunners fighting as infantry, disabled soldiers, militiamen, and armed government clerks; nor of Commodore John Randolph Tucker's Naval Brigade, sailors whose ships lie at the bottom of the James River. Both groups, in Ewell's view, doubtful in a fight.

He'd warned Lee that they bore close watching, particularly in a hard fight. Had he watched them this late afternoon, however, he might have worried less.

They're responding well to their officers' commands. Aware that across the valley, battery after battery of Federal artillery and regiment after regiment of Blue infantry are coming into line below the Hillsman farm, they don't seem unnerved at the sight but work harder on their breastworks.

The sun, already well down in the west, casts long shadows across the oak, pine, and dogwood crowning Ewell's hill; its fading rays laying an early evening grayness across the valley's boggy floor. Below Ewell's hastily built breastworks, scattered broom sedge and low pines offer excellent fields of fire for about three hundred yards.

At the bottom of the slope is Sayler's Creek, lined with willows, alders, locusts, and underbrush that is thick in places and crisscrossed with trees felled by past storms. A twisting, turning, flooded creek whose banks often are hidden by still-high water. Beyond the creek: soft, muddy, open bottom land that gradually slopes upward toward the Hillsman farm and the gathering Northern regiments. A rifleman, running across that bottom land, can stumble into the creek before he knows it's there; stumble into deep, shoulder-high pools. Hard to keep a decent skirmish line there; harder yet to be wounded there.

Federal infantry, Wright's VI Corps, mean to cross that valley. No doubt about that. Kershaw and Custis Lee, with their soldiers yards apart behind the Confederate breastworks and no reserves, wait to dispute their passage.

At sunset Wright, urged by Sheridan to hurry his attack, sends two Federal divisions into the valley. Meanwhile, five Union artillery batteries pound Ewell's lines. Gray soldiers huddle behind any shelter they can find as solid shot and explosive shells tear the ground, splinter trees, burst around them.

As Colonel Oliver Edwards' brigade of Massachusetts' soldiers march down into the valley, Sheridan gallops up to point to Ewell's breastworks on the opposite slope.

"Colonel, you see that enemy line over there? When you get across that creek, form your brigade in one line and carry those breastworks. Carry them! Do you understand me?"

"Yes, sir, but whose on my left and right, General?"

"Dammit, never mind your flanks, Colonel. Cut through their line, and your flanks'll take care of themselves. Why they're as demoralized as hell over there. You can break their whole line with your brigade alone. Get it done, Colonel."

As the long lines of Federal infantry move steadily into the valley every regiment loses soldiers to Confederate sharpshooters. Mounted officers are particularly vulnerable. When a man falls, Union infantrymen close the gaps in their ranks and keep going.

When they reach the creek, some soldiers stumble over its hidden banks into deep pools. Protected, for the moment, by the high bank, they struggle across the chest-deep, spring-cold water.

Once rifle squads are across the creek, sergeants re-form their lines and urge them up the hill. They've lost men to the sharpshooters; they still must face solid volleys of rifle fire from the Confederate breastworks.

A young Rhode Island lieutenant, nervously entering his first fight, crosses the rutted, dirt road above the creek, trying not to look at the dead already there as he steps around them. One lies beside the road, as if asleep, a neat, round hole through his forehead.

Then their company commander waves them forward.

As they climb the slope, he realizes that the company's veterans are half crouching as they advance, and he determines to set a better example by standing erect. Then he feels the "whiff" of a Minie ball past his right ear and instinctively ducks. He's gone another dozen steps before he realizes that he's now crouching just like his veterans.

A few steps more, and he suddenly feels a heavy, dull blow on his left hip, like being hit with an icy snowball. Not the sharp, burning pain he'd expected but more a dull, spreading pain. Then it becomes sharper, and he crumbles to the ground.

When they're within a hundred yards of the Confederate breastworks some Federal officers and soldiers begin to wave handkerchiefs.

A puzzled South Carolina private calls to his sergeant, "Them Yanks fixin' to surrender to usn's, Sarge?"

"Not likely, boy," the answer. "They mean for us to surrender to them. Figure we don't have a chance and want us to lay down our rifles. Good of 'em, but not likely. Get ready now."

Then their captain raises his arm and calls out, "Ready!"

Gray soldiers, like some terrible machine, rise together to one knee.

When he commands "Aim!" they level their rifles just above their enemies' knees.

"Fire!"

Their volley rips the first Federal line from one end to the other. When the cloud of gray-black smoke begins to lift, they see that the Yankee line has been cut down, as by a scythe. Some of their enemy are lying very still; many more writhe on the ground like snakes hoed down at the end of a furrow. Others stagger to their feet only to be cut down again.

A second Confederate volley, almost as deliberate as the first, and the Federal line, struck from one end to the other, breaks.

Breaks on Colonel Oliver Edwards' brigade; Edward's men finding out what they knew all along: Sheridan was wrong; there's no way they could cut through the Confederate line by themselves.

STILES

The Confederates on the left of Custis Lee's line are Major Robert Stiles' James River artillerymen fighting as infantrymen.

They're working on their breastworks when Commodore Tucker's marines and sailors move into position on their right.

"To the starboard, march!" Tucker calls, and Stiles' men smile at the sailors, "Aye, aye, sir!"

An infantry staff officer asks Tucker if he can help the commodore position his men. If Tucker recognizes the affront, he handles it well.

"Thank you, young man, but we'll need no help. My people understand my intent. Tell the general we'll do just fine as we are."

When the Federal artillery bombardment begins, Stiles concentrates on his own men, ordering them to hug the ground. He walks just back of his line, talking to them as quietly as he'd done at their religious services. The younger ones, nervous, frightened, are steadied by his voice. Stiles feels better about things; they'll fight when it's time.

A twenty-pound artillery shell strikes Stiles' line, tearing one of his artillerymen in half and flinging the soldiers' head and torso into the air. The corpse strikes Stiles who uses his ragged sleeve to wipe away blood and torn flesh.

The body, he recognizes, is that of Private Blount, the young soldier who'd been so deeply moved by Stiles' reading the Soldiers' Psalm the night their retreat began.

Rifle fire rips Stiles' breastworks, and he hears the "thunk" of a bullet striking a body. One of his men has been knocked to his knees and, clutching his neck, is screaming, "I'm killed! Killed!"

Stiles, there in two strides, finds a still-hot Minie ball, slowed by the man's blanket roll, shallowly lodged in the soldier's neck. He pulls it free and hands it to the soldier.

"Here!" he grins, "and you're not killed either. Just lucky. When you get home give it to your sweetheart."

The soldier grins and picks up his rifle.

At fifty yards Stiles sees Federal soldiers waving their handkerchiefs. He shakes his head, checks the loads in his revolver, and begins his commands.

Even Stiles isn't prepared for the effect of their volley, the first line of Blue infantry disappearing in the heavy smoke. When it begins to clear, few of their enemy are still standing. At their second volley the Union line breaks, its survivors falling back.

Then, without taking time to reload and without Stiles' command, his men scream their high-pitched Rebel yell, vault their barricade, and begin to chase their enemy down the slope.

"Halt! Halt! Come back!" Stiles yells, but this time they ignore his command. Determined to finish off their enemy, they rush down the slope after them.

Near the creek bank, as Captain Bradley, Adjutant of the 37th Massachusetts Infantry Regiment, grapples with a Southern officer, a nearby Confederate rifleman shoots Bradley in the thigh. Bradley, falling, clings to his enemy and together they roll down into the creek.

As they struggle, the Confederate officer also shoots Bradley. Then the Southerner staggers to his feet, aiming his revolver for a final shot. Before he can fire, however, a Massachusetts' rifleman, Private Samuel E. Eddy, fires his own musket, and the Confederate officer falls dead across Bradley.

Eddy, reloading his musket, in turn is bayoneted by another Confederate soldier.

He feels the bayonet's sharp, triangular-edged blade scrape his ribs then drive deep into his body. It's over in a second, but to Eddy, aware that the bayonet is being pushed through him, it seems an eternity. His knees buckle, and he feels the blade break through his back, near the spine. Pinned to the ground, he's conscious through it all, aware of what is happening to him.

His enemy, a youth no older than Eddy, braces his foot against Eddy's side to free the bayonet. Eddy, however, has clung to his own weapon and, with its muzzle pressed against his enemy's body, pulls its trigger.

The force of the bullet throws the Confederate soldier upward and away, dead before he hits the ground.

Grasping the fallen man's musket with both hands, with one convulsive jerk Eddy pulls its bayonet free.

Nauseous and in terrible pain, he lies quietly for a few moments, aware of the pain and that he's bleeding badly.

Then he staggers to his feet. Seeing that the melee has passed beyond him, and deciding that he's done all that he can do this day, he picks up his rifle, stumbles across the creek, and follows the bloody trail of other wounded to the dressing station at the Hillsman farm.

Meanwhile, Major Robert Stiles, his charging men ignoring his shouting, "Halt! Halt! Come back here!", shrugs his shoulders and runs down the slope after them. Near the creek he catches up with his color bearer, a young, quiet boy.

Spinning the youth about, Stiles demands, "What do you mean advancing the flag without orders?"

Before the boy can answer, a shell explodes nearby, and they're both thrown to the ground. Stiles staggers to his feet; the young color bearer is dead, his butternut jacket torn and already blood-soaked.

Another of Stiles' soldiers picks up the fallen flag.

"That's my brother, Major. The flag's mine to carry now."

Before he can move, however, he too is shot down.

Another man reaches, "Give 'em to me, Major," but then he too is shot down.

Beside Sayler's Creek Stiles sees a half dozen of his artillerymen lying dead or wounded beside their flag.

"I'll not lose another man over this flag," Stiles tells himself and, jamming its staff into the ground, he wades across the creek to find his battalion.

Huddled on the slope below the Hillsman farm, their impromptu charge has run its course. Moments before the next wave of Federal infantry appear, he's able to hustle them back to their breastworks.

This time Union soldiers seem to be everywhere, rank after rank of them, and they're not waving handkerchiefs. Many of them fall at Stiles' volleys, but the rest keep coming, tearing through his skimpy barricade then using revolvers and musket butts and bayonets to finish off his artillerymen turned infantry.

The fight becomes a confused melee of brutal, personal, one-on-one fights: men killing each other with bayonets, rifle butts, or knives. Some, without weapons, throw stones; bite each other's ears, and noses, and throats; roll on the ground for a critical advantage.

A Union captain and a muscular Confederate officer wrestle for Stiles' guidon, each man using the sword in his free hand to hack away at his enemy. The Confederate slashes his opponent but within seconds also falls, the disputed flag disappearing beneath a flood of Federal soldiers.

Stiles, slowly, deliberately firing his revolver and constantly glancing left and right at his line, sees one of his men kill another Confederate soldier, wearing a captured blue overcoat, mistaken for an enemy.

The fight has passed over them, and somehow Stiles has survived. His line broken and his men scattered, he turns to run, but not far; a squad of Yankee soldiers have come in from his rear. For Major Robert Stiles the fight is over. Exhausted, he's proud of the fight his men put up and that he's ended his own service to the Confederacy in battle.

TUCKER

Fighting to the right of Stiles' artillerymen is Commodore John Randolph Tucker's battalion of marines and sailors.

Massachusetts' soldiers approaching Tucker's breastworks can't figure their naval uniforms, one astonished Yankee exclaiming, "Damnation, have they got gun boats way up here?"

Tucker's men fight hard. They're fighting beside army men who've laughed at their uniforms and at their commodore's odd commands, and they won't give up until they've exhausted their ammunition then fought to the end with bayonet, musket butt, and navy cutlass.

In the thick of the fight, with Yankee soldiers everywhere and his position about to be overrun, Tucker grins and tells a subordinate, "I've never been in a fight like this before, Captain, but everything seems to be going well."

When his breastworks finally fall, he leads his survivors to a little draw where, overlooked by Federal soldiers, they hide until they're discovered much later in the evening. Even then they refuse to surrender until a staff officer comes with a direct order from Ewell for them to lay down their arms. They do so, reluctantly but with honor, knowing that no infantryman of either side has done better than they.

"It's ironic, General," Tucker will tell Ewell, "that my marines and sailors would have a little country stream called 'Sayler's Creek' to show their worth."

Sergeant Angus Cameron's squad of the 5th Wisconsin Infantry Regiment captures Ewell. Custis Lee, Kershaw, and five more Confederate generals also are taken. Pickett and Anderson, however, escape in the confusion following the collapse of their lines. Anderson has lost about 2,600 of his 6,300 men; Ewell 3,400 of his 3,600. And the fight still is not over; not for John Gordon's men, several miles away.

GORDON

About 4 p.m. Gordon arrives at the upper end of the valley to find that the wagon train he's been following has bogged down trying to cross two branches of flooded upper Sayler's Creek by flimsy, one-wagon-at-a-time bridges. As he's making some order of the confusion there, heavy firing erupts behind him. Union General Humphreys' II Corps' skirmishers are closing fast on the line Gordon had thrown across the Lockett farm, above the bridges, to protect their rear.

Gordon's artillerymen fire across the valley at Federal infantry advancing down the opposite slope, cheering as an entire team of Federal color bearers fall to their fire.

Despite their losses, Humphreys' men push Gordon's infantry back to the creek. There the fight becomes another hand-to-hand melee with Union and Confederate soldiers dodging in and out of the wagons stalled at the bridges, firing into each other. Seeing the massive numbers before them, Gordon's men abandon the wagons and fall back toward his hastily built breastworks.

Near the stream, a Confederate colonel, his leg shattered, has his men prop him against a large pine tree where he continues to fire at his enemies until, his revolver empty, he flings it on the ground and surrenders.

Another Gray officer, struck hard in the chest, finds that a spent musket ball has penetrated his blouse over his heart but been stopped by a spur he'd put in his pocket.

A third Confederate officer, near the abandoned wagons, yells to his men, "Clear out boys. Take care of yourselves." Then, planting himself against a tree, he deliberately empties his revolver at his attackers before running to join his men.

In the middle of the fight, a sergeant with a problem interrupts Gordon.

"General, I can't fight those Yankees and watch these here prisoners too. We've got our tails in a wringer here, sir. Reckon I should just turn 'em loose?"

The prisoners, the two Northern couriers captured in Confederate uniforms with Grant's message to Ord the night before, are scouts of Major Henry Young's detachment.

When Gordon's men first ordered them to throw down their weapons, they'd laughed, insisting that they were Confederate soldiers. Then they answered Gordon's questions easily, rattling off the names of Fitz Lee's units and their commanders.

They convinced Gordon; his sergeant, however, wasn't persuaded. He insisted that they search the prisoners' boots and, when they did, they found Grant's messages.

"Well," Gordon told them, "we took you in our uniform, and you know what that means. Sorry, but in the morning I expect we'll have to shoot you."

The elder of the two, still a teenager, spoke for both prisoners.

"Now, look, General, we knew what we were doing. Took our chances as any soldier would. And you have the right to shoot us. But what good would it do? This war's about over; what's to be gained by killing two more of us?"

Gordon thought about the problem, then referred it to Lee. Later in the day Lee sent back his answer.

"Keep them, General. Just keep them, until I've had time to think about it."

Now the sergeant, knowing their situation as well as Gordon, asks the same question they'd asked, "What good will shootin' two more of 'em do, General? They didn't mean no harm. Why don't we just send 'em over to the Yankee lines?"

"All right, Sergeant," Gordon decides. "Rig up a white flag, and send 'em over. Reckon they're two mighty lucky Yanks; maybe some of it'll rub off on us."

Later, he'll release another prisoner. In the middle of the fight a young Massachusetts' lieutenant waves a white flag before Gordon's line. When he's waved forward, he comes to Gordon, salutes, then reports.

"General, the Colonel's compliments, sir, and he's asked me to point out the hopelessness of your situation and beg you to lay down your arms before we have to attack."

John Gordon's deep-set, hawk-like eyes flash.

"Young man, I appreciate your Colonel's concern and your courage in coming into my lines this way, but you're mighty lucky that one of my men didn't shoot you. I hope you realize that. And you're making a worse mistake asking us to surrender. Go back to your friends, sir, and tell them that we will not surrender."

The fight to break John Gordon's lines is every bit as heavy as any part of Anderson's or Ewell's battles, but when dusk ends the fighting Gordon has stopped the Federal assaults before his breastworks. Union bugles sound "Recall," and Humphreys' soldiers leave the slope to their enemy.

Gordon gathers survivors from the scattered units. Most are from his own regiments, but many were with Ewell's and Anderson's corps, and they bring alarming news of entire divisions being overrun then taken prisoner. He leads them all toward the High Bridge crossing of the Appomattox.

Gordon's own losses have been heavy: several thousand men killed or wounded and another 1,700 taken prisoner; and three cannon, 200 wagons, 70 ambulances, and 13 battle flags also lost.

He's done the best he can, however, and his report to Lee is restrained:

> I have been fighting heavily all day. My loss is considerable and I am still closely pressed....So far I have been able to protect (the trains)...but without assistance can scarcely hope to do so much longer. The enemy's loss has been very heavy.

LEE

Lee, anxious about reports that Gordon is heavily engaged and over the lack of word from Ewell or Anderson, has ordered Mahone's infantry division back to Sayler's Creek. He'll go with them to see for himself.

North of Rice he rides to a hill, dismounts, and holding Traveller's reins in one hand, sweeps the hills and valley before him with his field glasses.

There are some white objects he can't quite make out in the distant valley, and he motions to a nearby officer.

"Captain, you have a young man's eyes. Those white objects," he points, "are they sheep?"

The officer shades his eyes with his hands, studies the valley, then shakes his head.

"No, sir, they're not. General, I think those are Yankee wagons."

Lee again raises his field glasses then slowly nods his head.

"Yes, you are right. But what are they doing there?"

As if to himself he adds, "And why haven't I heard from Ewell and Anderson? It's strange that I've not heard from them."

He turns to Mahone, "General, please move your division forward."

As Mahone's infantrymen near the slope above Sayler's Creek, Lee and Mahone ride ahead for a clearer look at the valley.

At his first view Lee's gauntleted hand nearly drops his field glasses. Feeling the slack rein, Traveller stops. Then Lee, standing erect in his stirrups for a better view, sees something that he never expected to see. The army, *his* army, in total, disorganized retreat. Routed.

Ahead of them, horses gallop without riders. Wagons are burning. Cannon lie abandoned. Soldiers, glancing over their shoulders as they half-run, half-walk toward them, are hurrying up the road. Many have no weapons. Behind them, a steady stream of wounded limp toward them. All fleeing whatever happened to them in the valley.

In the distance clumps of Union soldiers seem to have broken through hasty breastworks across the valley's near slope, and bugles, dozens of bugles, faint but insistent, are calling more Union troops up from the valley's floor. He can't identify any large Confederate units still intact.

"General Mahone," Lee guesses, "I believe that there may have been two separate fights, about a mile apart. And a mile beyond them, there is more fire and smoke. Another fight? Whose commands? And what has happened to Anderson and Ewell? Surely two whole corps couldn't have vanished?"

Then, dropping his field glasses to his side, for once he loses his discipline to exclaim, "My God! Has the entire army been dissolved?"

Mahone, as alarmed by Lee's reaction as by the scene before them, tries to reassure him.

"Not the whole army, General; not the whole army. My men are ready to do their duty."

Lee breathes deeply, recovers his poise, scans the frightening scene again then turns to Mahone.

"Yes, I know; and I'm grateful. General, please keep those people back while I see to the remainder of our army."

When Mahone has gone to hurry his men forward, Lee looks about, searching for something that *he* can do. Then he motions to a nearby color bearer, standing with a blood-red and blue and white Confederate battle flag.

Taking it from him, and holding his hat in one hand with the flag in the other, Lee simply raises the banner above his head.

At first Traveller is skittish about the red folds of bunting flapping about him; then he stands absolutely still. Horse and rider are unmistakeable to the men hurrying up the road toward them.

Some of Lee's retreating soldiers, confused, uncertain, frightened, stare at him as if blaming Lee for whatever happened in the valley behind them. Then they hurry by. Most of the demoralized men, however, steal embarrassed, humiliated glances at the lone horseman on the big, gray horse and begin to flock around him, as if immediately finding strength and reassurance in the calm that Lee radiates.

Mahone returns to find Lee's magic at work. Officers and sergeants are reorganizing units, persuading beaten soldiers to turn about and face their enemy. Men are gathering weapons, redistributing ammunition, and searching for their regiments.

Mahone smiles and takes the flag from Lee's hand. Lee, calm now, nods and again studies the valley before them.

Then, using Lee's worn map, the two generals work out a plan. Lee will return to Rice and Longstreet's corps. Mahone will delay any Federal pursuit then, covered by the night, retreat across High Bridge to Farmville. He'll also send riders to point Anderson's, Ewell's, and Gordon's survivors toward the bridge. When the last of the army have crossed High Bridge, Mahone's to destroy it.

———

At Rice, Lee learns that Gordon's men, separated from Anderson's and Ewell's commands, still are heavily engaged. Gordon, however, hopes to break away, cross the Appomattox at High Bridge, then rejoin the rest of the army near Farmville.

Scouts sent to find Anderson and Ewell haven't yet returned.

Incredible to think that Ewell's and Anderson's entire commands could be gone, but clearly that now is a very real possibility.

If so, he may have lost more than a quarter of his remaining army this day. That would leave him with only six divisions capable of putting up a hard fight, and only two of them of respectable size. His cavalry, Fitz Lee assures him, still are full of fight, but their horses are tired and dying. His artillery, half its gunners gone, have lost most of their guns and ammunition.

If he must fight in the morning, he mentally sums it, he'll have perhaps 12,000 reliable infantrymen and 3,000 cavalry to meet Grant's four corps of infantry and four divisions of cavalry, perhaps 80,000 Union soldiers.

Very bleak, but he'll not give up. Brigadier General Isaac St. John, his Commissary Officer, tells him that he has 80,000 rations of meat and 40,000 rations of bread waiting at Farmville, five miles above Rice. If they can get to Farmville by early morning, he may be able to feed his army and recross the river before Grant can reach the town. Then, protected by the Appomattox, they can steal a few hours' rest before continuing west to regain the lead on their pursuers, and this time hold it. Then swing south for Danville or, if that's not possible, march for Lynchburg and the Blue Ridge beyond. A gamble, but Lee has always been a gambler. The alternative, surrender, is not yet an option he'll consider. Not so long as they still can fight.

He issues the necessary orders.

———

Later in the evening Lieutenant John Wise, having ridden all day, finds Lee before a campfire outside his headquarters' tent.

Lee studies Jefferson Davis' telegram instructing Wise to determine Lee's intentions.

"You've had a difficult and dangerous journey, Lieutenant Wise. Your father will be proud of you. You've heard from him?"

"I've not seen him, General, and I understand that his brigade was heavily engaged in the fighting today."

"Yes. A disastrous action. Many of our units were overrun, and we still are not sure of their status. But I've been told that General Wise is safe and hopes to rejoin us at Farmville in the morning."

"Please tell President Davis," he continues, "that I've sent no written dispatches because I feared that they might be captured by enemy patrols. Our situation is very uncertain. I've not been able to strike for Danville. Because the rations I'd ordered were not at Amelia Court House, we had to delay our march. Now I must hold to the railroad until we can be supplied. If we can feed our soldiers at Farmville, we'll continue west until we can get ahead of our enemy. Then we'll march for Danville or, if that is not possible, strike for the mountains beyond Lynchburg."

"Have you chosen a place to make a stand, sir?"

Wise will always remember a momentary flash in the General's eyes, and almost a smile, at his question.

"No. That isn't possible. Not yet at least. Tell President Davis that I must wait for developments."

Then, almost as an aside, he's very candid, "Another Sayler's Creek, Lieutenant Wise, and it will all be over; ended. Just as I expected it would end from the start."

"Now," his gentle smile returns, "perhaps Colonel Marshall can find something for you to eat. Then you may wish to find your father before you return to Danville."

THE BATTLEFIELD

At Sayler's Creek darkness is falling fast as Union soldiers mop up scattered bits of Confederate regiments, see to the dead and wounded, and herd prisoners together for the march back to City Point.

Once the fighting ends and they've seen the faces of the ragged, hungry men they've beaten, Northern soldiers share their food with their prisoners. The blood lust has passed. Both armies have known heavy fighting and tasted defeat; so there is understanding and compassion.

A Union soldier finds a mortally wounded Confederate infantryman, lying on a bloody bed of leaves.

"What can I do for you, Johnny? I want to help if I can."

"Thanks, Yank; but no one can help me now."

He nods and has turned to leave when the Confederate calls after him.

"Yank, you might pray for me before you go."

The Union soldier has never prayed aloud. He wants to now, but he just can't do it. In desperation he calls to his squad.

"Come here, boys. Quick now. Here's a poor Johnny all shot to hell. He's dying and wants me to pray for him, but you know that I can't pray worth a damn. One of you must do it for me."

One does, slowly, hesitantly, and the hardest of them bows his head before they turn away.

There are worse sights: bodies of men, and horses, and mules. Alone or heaped together; twisted, torn, some seeming at peace while others seem still tormented by what has happened to them. Everywhere.

Hundreds of campfires have sprung up across the valley; survivors cooking their remaining rations.

Guided by the campfires and by the lanterns of medical teams crossing and recrossing the battleground, a steady stream of wounded, Blue and Gray alike, make their way to the Hillsman farm.

Stretcher bearers search the battlefield, making life and death decisions without batting an eye.

"Take this one; he's got a chance if we can get him back there before he dies. Never mind those two over there."

"Can you make it up there on your own, Johnny? There's others need us more."

"Don't worry about no guards. You're not goin' anywhere and, if you did, wouldn't bother us none. Just follow that path 'cross the creek and up to that farmhouse yonder."

Where the creek crosses the road, a deep ditch is filled with Confederate and Union dead.

A Union soldier coming upon that awful ditch stops, stunned by the sight.

A lone Confederate soldier is kneeling before it, his hands uplifted as if in prayer, his eyes open, and a calm smile on his handsome face.

The Union soldier, not sure whether his enemy is praying for a lost friend or for the horrible things they've all seen and done this day, touches the man's shoulder in sympathy. A slight touch but enough: the man falls forward, dead.

A bit farther downstream, during the desperate fighting near the creek's bank, a Union corporal had shot a Confederate officer during the Southerners' charge into the valley. It's the first time the Yankee soldier *knows* that it's his bullet that has struck another human being.

He lays aside his rifle and, ignoring the fight raging around him, sits in the gathering dusk beside the slowly dying man; deeply troubled but not knowing what to say.

Finally he confesses, "I'm the one who shot you, Johnny; and I'm sorry. Guess I was too scared to think about it, but I'm a Christian, and I'm sorry. If it's all right, I'll pray for you, for both of us. It's all I can do now."

The Southern officer, unable to speak, nods, and the young Massachusetts' soldier prays over him. When he's finished, he sits for a long time, eyes closed, holding the man's hand.

Then he feels the pressure on his fingers ease. When he opens his eyes, the man is dead. The corporal gently closes the eyes of the man he's killed, stands in thoughtful silence, then takes up his rifle. Best he go find the rest of his squad somewhere up that hill.

The hillside is strewn with the debris of battle: bits of clothing, canteens, blanket rolls, hastily discarded playing cards, tents, kettles, and desks. It also is white with Adjutant General's papers, commissary returns, and other paperwork of an army. Dotted too with hundreds of soldiers' little Dutch ovens, and with the bodies of the men who'd used them.

The devastated Confederate wagon train is of particular interest to Union soldiers. Several hundred Confederate army wagons block the road. Many of their canvas tops bear chalked messages from their previous owners: "In the last ditch!"; "The C. S. A. is all gone up!"; "We-all can't beat you-all without somethin' to eat!"

Many of the abandoned wagons bear the symbols of their military units, some Northern, some Southern. Northern officers with a mind for history are thankful for their victory but saddened at the reminders of legendary Gray units: Stonewall Jackson's old brigade, the Louisiana Tigers— now all gone.

Officers' camp equipment: swords, trunks, barrels of apple jack, toilet articles, and other fancies are everywhere. In one wagon a Union soldier finds a half-dozen fat puppies. In another, they find General Ewell's personal uniforms which a Yankee drummer boy, Delavan S. Miller, promptly dons then requires his messmates to salute him.

They also find a Confederate paymaster's wagon with more than $400,000 in crisp new Confederate dollars.

"Hey, Johnny," a Yankee scavenger calls to a group of dejected Gray prisoners, "when did you last draw your pay?"

"Hmph, about as long ago as you think, Yank. It was so long ago that I disremember. Why, you gonna pay us?"

"Sure. Here you are," and laughing he begins to throw bundles of Confederate bills from the paymaster's trunks.

Later, Federal campfires become miniature Monte Carlos with $10,000 in Confederate dollars the usual opening bid; another $20,000 to raise the ante; and perhaps another $50,000 to sweeten the pot.

Confederate Major Edward M. Boykin looks in amazement at nearly 900 Federal prisoners being herded across High Bridge. The prisoners' uniforms seem so new, so clean, with shining buttons. And Grant's soldiers seem big, well fed.

A large pile of captured Spencer repeating rifles have been loaded in wagons. Boykin takes one for his own use, stuffing cartridges in his coat pocket.

That night he'll share his blanket with a Confederate colonel and a lieutenant, survivors of Pickett's fight at Sayler's Creek. Their regiment, their entire division, legends after Gettysburg, now gone.

Confederate drummer boy John L. J. Woods was with his regiment as it climbed the south slope above Sayler's Creek. Then he'd beaten the Long Roll calling its soldiers to arms.

Because Johnny is so young, however, and perhaps sensing what is about to happen to them, his Sergeant Major sent him to the rear to help the regimental surgeon.

While he's waiting for wounded to come in, young Johnny Woods wanders off to find food.

Before he can return, Federal cavalry cut the road behind him and a farmer advises, "You'd best skedaddle, boy, while you can. Get over that long bridge up there a mile or two. Lest you hurry, you're gonna get cut off too."

He crosses the bridge with Gordon's men. That night he'll sleep under a tree. Beside him, as he sleeps, is his drum. Johnny Woods: a veteran.

SHERIDAN'S CAMP

Sheridan camps not far from Sayler's Creek, an unpretentious camp of a half dozen tents and as many campfires. In a nearby orchard, Custer's cook, Eliza, is preparing a meal for Sheridan, his generals, and their prisoner-guests.

Sheridan is lying with his back on a saddle blanket, his feet to the fire, a smile on his wind-burned face. He's sleepy but still awake. A number of other officers are huddled around the campfire: some in blue uniforms, some in gray. Most of the ranking captured Confederate officers are there.

Ewell sits on the ground, hugging his knees, staring into the fire. He hasn't much to say, but he shares with Union General Horatio Wright, whose blankets he'll share this night, "Our cause is gone, General. Gone. Lee should surrender now, before more lives are wasted."

Prisoner or not, he performs one more duty before sleeping, sending for Major John Stiles and, before Sheridan's officers, congratulating him on the conduct of his battalion during the fight.

Custer arrives, a long day in the saddle and a terrible fight behind him but still a bundle of nervous energy. Custer, standing out among the others, even now, in olive corduroy blouse and trousers, a navy shirt with large silver stars on each wide collar, and a flaming-red flannel neckerchief bound by a gold "CUSTER" pin. His long, yellow hair reaches his shoulders.

South Carolina General Joseph Kershaw and Colonel Frank Huger, an old friend from Custer's West Point days, will share Custer's blankets this night.

"Hello, General," Custer teases Kershaw. "Glad to see you. We've met so often that I feel as if I ought to know you."

"Yes," Kershaw smiles back. "We've met often in the past, but I'm afraid not in the best circumstances for cultivating an acquaintance."

Blue and Gray officers share coffee, sugar, condensed milk, hardtack, and boiled ham on Eliza's tent fly table cloth. Steak and onions to hungry men. Then they smoke as they talk around their fires, dwelling on the past as the future for Custer's guests is not inviting.

A short night then, long before dawn, Custer is up, preparing to take up the chase.

GRANT

Before he sleeps, Sheridan sends a messenger to Grant:

...we have captured Generals Ewell, Kershaw, Barton, Corse, Hunton, Dubose and Custis Lee; several thousand prisoners; 14 pieces of artillery; and a large number of wagons. If the thing is pressed, I think Lee will surrender.

Grant relays Sheridan's message to Lincoln at City Point and scribbles a note to Meade reminding the still-ailing general that "Every moment is now important to us."

Then he orders Sheridan and Ord to block Lee's retreat at Farmville while Meade attacks the Appomattox crossings at High Bridge.

Griffin, whose V Corps did not see the fighting at Sayler's Creek, instead marching thirty-two miles while picking up hundreds of Lee's stragglers and their equipment, bivouacs near the battlefield.

Griffin, like all Grant's other commanders, is short of supplies. Everyone, however, is elated at the victory and anxious to finish off Lee's army by a hard march or a hard fight, whichever it takes.

"In the morning," Grant smokes his last cigar of the day and decides, "Griffin can swing down to the left behind Ord in case Sheridan needs help. He did a long march today, and it'll be another tomorrow, but Griffin can do it." He smiles to himself and finishes the thought, "If he puts the Professor in the lead, he can do it."

He finishes his order: all the troops are to march as soon as possible.

LINCOLN

Back at City Point, Lincoln has welcomed Mary Lincoln, back from Washington with a mixed party: a young French count, Adolphe de Chambrun, a nobleman with impeccable manners who, to Lincoln's amusement, practices subtle flatteries on Mary Lincoln; Lincoln's old political enemy, Senator Charles Sumner, who'd insisted on coming along to see the captured Rebel capital; James H. Harlan, his Secretary of the Interior, and Harlan's entire family; and Mrs. Lincoln's maid, Elizabeth Keckley, who'd once been Varina Davis' seamstress.

Senator Harlan, an old friend, is amazed at his first glimpse of Lincoln. The President's poise and bearing, in fact his entire appearance, have miraculously improved during his few days at City Point. The sadness he'd come to expect as a permanent feature of Lincoln's countenance is gone. Instead Lincoln seems very happy and very much at peace, as if conscious that he finally is achieving the great purpose of his life.

Mary Lincoln fails to see that miraculous change or, if she sees it, does not approve of it. She's unhappy that he'd gone to Richmond without her and promptly announces that tomorrow she'll go there herself.

He doesn't want to return to Richmond, suggesting instead that he host a tour of Petersburg. She quickly rejects the idea, insisting that Richmond is her choice. Lincoln's countenance, Harlan sees, already is losing its joy.

That evening Lincoln telegraphs Grant, advising him that because Secretary Seward had been badly injured in a carriage accident the Lincolns must return to Washington.

Then he congratulates Grant on his great victory at Sayler's Creek and closes, "Sherman says, 'If the thing is pressed, I think that Lee will surrender.' General, let the *thing* be pressed."

CHAPTER 5

Let the Thing Be Pressed   Good Friday, April 7, 1865

LEE

He hasn't slept. Not this night. Exhausted, he dozes in his camp chair, but he doesn't sleep. Couriers come and go. Gordon has broken free of his pursuers, gathered fragments of Anderson's and Ewell's scattered commands, and is crossing High Bridge. Behind him, when Gordon's men are all safely across, Mahone's rear guard will fire High Bridge and a small wagon track bridge spanning the Appomattox below it. Destroying the bridges will considerably delay Meade's pursuing infantry.

Mahone's a reliable general; still Lee will feel better when the bridges are down.

The pain across his chest is greater at night. "I must stay calm," he counsels himself, "that seems to ease it." He knows that his staff worry about him; twice, when he'd raised the tent flap to glance at the dark sky, Colonel Long by the fire had been instantly concerned until he'd smiled his thanks and turned back.

Snow. A late spring squall; not heavy, and maybe it will help the roads ahead hold up a bit longer. But cold, cold for his soldiers. Well, at least he can feed them. Longstreet's scouts, already in Farmville, have confirmed that the rations are there.

Long before dawn he gives up on the idea of rest, mounts Traveller, and with a few of his staff reaches Farmville at sunrise. He finds that Longstreet's already picketed the town and is distributing rations to his men. The word will spread to new units as they arrive.

Lee, suddenly very tired, allows Colonel Marshall to persuade him to rest for two hours in Patrick Jackson's Farmville home and accepts a cup of make-do coffee but declines the food that is offered. Not yet, not yet.

Brigadier General Isaac St. John, his Commissary General, reports that several trains with more provisions are expected at Appomattox Station, some twenty-five miles up the Southside line toward Lynchburg. The ration trains will wait there for Lee's orders.

John C. Breckenridge, Davis' Secretary of War, en route from Richmond to Danville, unexpectedly arrives, and the two men confer. Lee has lost more than a quarter of his army at Sayler's Creek, but many Confederate units have survived the fight. If the pursuing Federals can be delayed long enough for him to feed his soldiers then recross the river and burn the Farmville bridges behind him, he can rest on its north bank. Then their outlook will be better. Still, his army is in great danger, and he shares that with Davis' emissary.

"General Breckenridge, you see the condition of my army. Our survival depends upon the legs of our soldiers. If we can regain the lead and feed our men, perhaps at Appomattox Station, we may yet be able to march for Danville. If not, we'll strike for the Blue Ridge. If we can survive the next few days, there is hope. Report that to President Davis."

"You've seen the proclamation he issued at Danville?"

"Yes," Lee smiles. "Mr. Davis has always been more sanguine than I about this war." He pauses then extends his hand, "Tell him that we'll do the best we can. Now I must see to my army."

He finds time to call on the widow of Confederate Colonel John Thornton, killed two years before. Then he returns to the depot where St. John's quartermasters are busy issuing two days' rations to Longstreet's men and to Fitz Lee's cavalry.

CLAIBORNE

Doctor John Claiborne is there, noting that the Lee he sees in Farmville seems as calm, confident, and incapable of fatigue as he'd always seemed riding the streets of Petersburg, it seems so long ago.

Lee, on the other hand, might not have recognized Claiborne. The doctor, wearing the same clothes he's worn since leaving Petersburg, is unshaven, dirty, and very tired.

Yesterday, Claiborne was scouting ahead of his little medical caravan when he'd encountered a North Carolina quartermaster who had half a canteen of whiskey hidden beneath his wagon's seat. He offered the doctor a pull of the canteen, and Claiborne had taken his full share.

Until then he'd held up well but, as the fiery liquid made its way to his empty stomach, he suddenly felt very hungry, tired, and afraid. He walked a mile or so beside his horse to stay awake but, when he reached Farmville, he tied the horse's reins to his wrist and fell asleep on the sidewalk. Then he'd awakened, cold, at dawn to see the sidewalk covered with others who could go no farther.

He finds that there's little ceremony in drawing rations. St. John's quartermasters simply put barrels of hardtack and meat along the sidewalks, and soldiers spear sides of meat on their bayonets, fill their pockets with biscuits, then hurry across the railroad bridge north of town. There's no paper work involved in drawing rations for whole units, wagon drivers simply carrying off whatever they can wangle from the clerks. Claiborne talks a corporal out of a side of middling meat and other provisions.

LEE

Below Farmville, a Federal courier with a white flag signals pickets of the 35th Virginia Cavalry Regiment.

"I have a message, Captain," he calls. "From General Grant for General Lee. Is he here?"

"That's none of your business, Major. You know that. And I won't accept a truce so long as that infantry behind you keeps advancing. Stop them and I'll see that General Lee gets the message."

The Union officer halts the skirmishers who've been pressing Longstreet's pickets. Then, having made his point with the Yankee major, Captain Frank Myers accepts the message.

Good news for Robert E. Lee: his son, Custis, is a Federal prisoner but he's alive, unwounded, and well. No word of Anderson, Ewell, or Pickett, but Custis is all right. Lee is grateful for Grant's thoughtfulness.

For a long time he's tried to picture Grant. He recalls meeting him once before, nearly thirty years ago in Mexico. They'd called him "Sam" then. Having fought Grant for more than a year, he knows the man well; but Robert E. Lee can't recall what Grant looks like, and that troubles him. More important, however, Grant's sending word about Custis reveals a sensitive, considerate man; something for Lee to tuck away in his mind.

Grant's message reaches him near the Farmville railroad bridge where his officers have persuaded him to accept a slice of bread and a leg of fried chicken.

Then, among passing soldiers, Lee sees a young nephew, George Taylor Lee. He'd not known until then that young Lee, a Virginia Military Institute cadet, had left his studies to join the Confederate army before Richmond.

"My boy," Lee smiles and offers his hand, "why did you come here?"

"I thought it my duty, sir."

"Son, it's good that your duty is so important to you. You're a good soldier. But you shouldn't have come on this long march," the general shakes his head. "There is little you can do here."

"If I'd stayed home, sir, I'd be a Yankee prisoner."

"No," Lee smiles, "I don't think they'd bother you; you're too young."

"Is Custis," the boy's favorite uncle, "here, sir?"

"No. I'm afraid that he was captured yesterday, but General Grant has assured me that he's all right. Thank you for asking. Now, have you had any breakfast?"

"No, sir."

Handing the boy his own untouched bread and chicken and gently patting the cadet's shoulders, Lee tells him, "Go somewhere now, George, and eat this. I must attend to other matters."

As Lee rides northeast toward High Bridge he encounters survivors of the Sayler's Creek fight. Among them is General Bushrod Johnson, whom he's been told fled the battlefield when his division was overrun.

Johnson seems uneasy and uncertain in his report, "We held as long as we could, General; but we were forced from the field. I've lost my entire division, sir, destroyed."

Even as Johnson speaks, Lee knows that his report is not correct because, farther up the column, he sees officers and men he knows are part of Wise's brigade of Johnson's division. Either the general doesn't know his own command or he failed to search it out after the battle. Either inexcusable to Robert E. Lee.

"No," he shakes his head, "some have survived. I see General Wise's men."

Johnson, turning in his saddle, stares in surprise at the approaching flags.

"There is General Wise," Lee points, "though," he smiles, "I hardly recognize him in his war paint."

Wise, recognizing Lee, but for the moment not noticing Johnson, salutes.

"Good morning, General. I report with the division General Johnson once commanded, sir."

Recognizing the not too subtle message and struck by Wise's appearance, Lee returns Wise's salute but pauses a moment before replying.

Brigadier General Henry Wise, formerly the governor of Virginia, lost all his personal baggage at Sayler's Creek, and he's been through a desperate fight then a hard night's march. He's very hungry, very tired, very angry.

At dawn he'd washed his face in a puddle of water beside the road; water, like Virginia soil, as red as brick dust. Then Wise, having no handkerchief or towel, let the water dry on his face, leaving his features as red as an Indian's.

And during the night some solicitous soldier, seeing the elderly general asleep beside the road, draped a ragged gray blanket around him, fastening it around Wise's neck with a pin.

Muddy, red-faced, draped in the old blanket, the former governor looks more like an ancient Sioux warrior than one of Lee's generals. And to top it off, someone had given the hatless Wise a battered Tyrolean cap which he now wears.

"Well, General," Lee smiles, "I see that *you* at least have not given up the fight. You have your war paint on this morning. And how is your command, sir?"

"Fit for a dress parade, General; fit for a dress parade."

Then, seeing Bushrod Johnson among Lee's staff officers, the irascible Wise loses his sense of humor.

"General Lee," in his anger forgetting his comic appearance, Wise draws himself to Attention beside Traveller, "these men, *my* men, will not move another inch. Not another inch, sir, until they have something more to eat than the parched corn they've been forced to steal from their horses. Do you understand, sir?"

Lee smiles, "I do understand, General. And you are right. They deserve something to eat, and they shall have it. Have them move into that field. Colonel Marshall, please send for food for these men. Let them draw their rations right here."

As the men march into the field, Lee continues.

"Colonel Marshall, please find a uniform for General Wise. General, I want you to take command of all the men you can gather here from General Ewell's and General Anderson's commands."

"General Lee," Wise protests, "I can't handle a command that size. I haven't even a horse."

"Find one, General, or let Colonel Marshall know that you cannot find one and he will find one for you. Take command of all these stragglers, sir."

"It will ruin my brigade to take in these extra men."

"You must obey orders, General," Lee is firm. "Please don't make my job more difficult."

"General, I *shall* obey orders, or die trying. That is a true soldier's duty. But," he looks directly at Bushrod Johnson, "I must understand them. Your army is not being disorganized by the men who are deserting its ranks, sir, but by those *officers* who have deserted their men. Now, do you mean that I am to take command of stragglers of all ranks?"

Lee, an old soldier, hiding his smile behind his gloved hand, avoids the challenge.

"General Wise," his voice is soft, reassuring, but firm. "Find a uniform; wash your face; get a horse. Eat. Rest. Then do your duty. You must help me, General."

Wise draws the ragged blanket closer around his shoulders, stands a bit taller, and salutes. Then, casting another long, angry look at Bushrod Johnson, he stalks off to join his men.

A bit later Lieutenant John Wise finds his father in Farmville, sleeping on the ground among his soldiers.

When he wakens him the general exclaims, "Great Jehosophat, boy. I didn't expect to see you here. I thought that you, at least, were safe."

When the lieutenant has explained his mission and asked whether Wise has seen Lee, his father erupts again.

"Seen him? Not two hours ago, when I'd washed my face in a mud puddle and was wearing a blanket for a coat, we saw each other. And he laughed. Laughed, sir. Told me to go wash my face! Come, we'll find him together. I've thought of some things that I want to say to him."

Farmville's streets are filled with Confederate soldiers: men searching for their messmates, their companies, their regiments. Snatching food from the quartermasters' way stations then, disregarding their sergeants urging them to hurry across the river, stopping to eat it on the spot. Teamsters driving wagons, seemingly without purpose.

They find Lee on the back porch of a Farmville home, drying his beard over a tin basin of water. The older Wise completes his report of his part of the battle at Sayler's Creek then launches into another angry attack on General Bushrod Johnson and those like him.

"Skedaddled, sir, skedaddled. No other word for it. He up and skedaddled and left us there to cut our own way out. Then he came right here, tail between his legs, to tell you that he'd done his best but that we'd all been lost. It's not hard to lose your men, sir, when you're looking over your shoulder. Over your shoulder! Not a soldier's view, sir, let alone a general's!"

"Now, General," Lee calms him. "I understand and I will take care of that in time. But," he smiles reassuringly, "tell me what you think of our present situation."

"Sir, we have no situation. Nothing remains but to put our poor men on their poor mules and send them home in time for their spring ploughing. There's nothing else we can do. We've done all we can. This army, General, is hopelessly whipped. It's already endured more than I ever thought flesh and blood could take. And, General Lee," his voice rises as he again erupts, "the blood of every man killed from here on is on your own head."

Lee again raises his hand to calm the older man.

"Now, now, General. That's enough wild talk. Don't do it; not to me nor to your officers. My burdens are heavy enough without that. If I disbanded this army, what would our country think?"

"Sir, the country be damned!" Wise will not be calmed.

Then, as if patiently explaining to a younger man who simply won't grasp the facts, Wise continues, "General Lee, there *is* no country. There has been none for many months. *You're* the country to these men. *You*, sir. They have fought for *you*, without food or pay or clothes or care of any sort for more than a year. *You* are all they have left. And there still are thousands of us who will die for *you*. And your loyalty, General Lee, must be to this army, not to a country that no longer exists."

Lee doesn't answer, only nods and continues gazing across the fields where his men are cooking their rations.

Then he turns to the younger Wise.

"Lieutenant Wise, I'll give you a note for President Davis. An hour ago I met with Secretary Breckinridge, and he already has left for Danville with his own report. You must go there now, but be very careful. There are strong enemy patrols south of us. Take the north side of the Appomattox to a ford eight miles west of us; then swing down toward Danville. Be careful; Sheridan's scouts may already have reached the ford. Now take a moment to bid your father good-bye. I need him here with my army."

As Lee is completing his note to Jefferson Davis a courier arrives with shattering news. General Mahone failed to burn High Bridge and the wagon bridge below it, and Humphreys' Federal infantry and some cavalry are pressing Mahone and Gordon there. Meanwhile, other Federal forces are coming up the Rice Road below Farmville.

Mahone had charged Colonel T. M. R. Talcott with burning the two critical bridges. Talcott soaked their timbers with oil and assigned men to torch them. Mahone, however, had ordered that the bridges not be set afire until he personally gave the final order.

When it's time, Talcott can't find him. Near dawn, long after the bridges should have been burned, they find Mahone in Farmville. Then they set the bridges afire. High Bridge burns quickly. The low wagon bridge at its base, however, is of hard wood, slow to burn. And Humphreys' infantry are closing fast on both bridges.

Despite galling fire from Talcott's soldiers, Union infantrymen put out the High Bridge fire with water from fire barrels forgotten along its tracks.

Below High Bridge, soldiers of the 19th Maine Infantry Regiment scoop up water from the swollen Appomattox to douse the burning wagon bridge.

Four wooden spans at the western end of High Bridge have dropped into the river, but its brick piers survive and the damaged area can be planked. The wagon bridge, however, still holds, and Humphreys' soldiers already are pouring across it. Within a half hour the barrier Lee hoped would give his army several hours' rest no longer exists and, if he does not hurry, his entire army may be caught between Humphreys' infantry above the river and Ord's beneath it.

———————

Lee calls Brigadier General Porter Alexander, an outstanding young artillery commander, to his temporary headquarters.

"General, three miles above us," he points to his map, "is the road junction at Cumberland Church. We'll turn west there. East of that cross-roads those people are pressing General Mahone. Take every battery you can spare, and help him hold them back. Keep that road junction clear for our soldiers."

"Leave batteries here," he continues, "to support our infantry if they're attacked while trying to cross the river. When they've crossed, destroy the Farmville bridges. Destroy them; we dare not have another failure."

Alexander nods but pauses to study the map another moment.

Then he suggests, "Sir, I don't know this country, but the route you have chosen swings north then west. If we stay on this side of the river, won't we save about twenty miles?"

Lee studies the map again, but tired and harried and warned of more Northern infantry on the Rice Road below the river, delays his decision.

"You may be right, General. But first we must put this river between us and the enemy coming from Rice. Then we can dispose of their infantry above the river. We can't handle two enemy forces at one time. After we've stabilized our situation, we can consider our final route. Do as I have directed you."

As Alexander hurries to his horse he hears gunfire already crackling below Farmville: Fitz Lee's cavalry and Longstreet's infantry delaying Ord's infantry.

———————

Major John Boykin's South Carolina cavalry are with Fitz Lee's squadrons at Farmville.

As they trot past a boarding school for young ladies, pretty girls wave handkerchiefs and cheer Boykin's men.

His cavalrymen, aged and hardened by the war but still young men who should have been in school themselves, answer fluttering handkerchiefs with waves of their hands, touch their caps in salute, try to sit taller, straighter in their saddles. Every one of them, however, is painfully aware that their innocent admirers see them as brave young warriors galloping forward to save the town from barbarians, while the truth is that within an hour or two they'll have to abandon the town. Leave Farmville and its people to Yankee soldiers. Not a prospect they like.

They fight hard but then are pushed back, just as they feared, retreating across the railroad bridge just before Alexander sets it afire.

———————

Farmville is lost, with less than half of Lee's army fed. The ration train, still not emptied, is forced to flee west. And the Confederate army must take up its march again, without rest. North to small, white-pillared, wood-frame Cumberland Church then west; if Alexander's artillery and Mahone's infantry can hold that door open long enough for the Confederate army to pass.

CUMBERLAND CHURCH

As Humphreys' soldiers near Cumberland Church from the east Confederate artillery suddenly rake advancing Union infantrymen with solid shot, explosive shell, and canister. The fire catches Major General Nelson Miles' division in a long, dense, four-man abreast column jammed into narrow Jamestown Road.

Miles' leading ranks are cut down as if by a scythe; bodies torn to pieces and scattered across the road and nearby fields. The rest of his column is stopped dead in its tracks. Miles, expecting an easy fight, realizes that it isn't going to happen and waves his regiments into two long battle lines.

At first the Union attack goes well. A Confederate artillery battery is overrun and its cannon captured.

Brigadier General Bryan Grimes' North Carolinians, however, counterattack. In a violent hand-to-hand fight, Grimes' men retake their guns, double load them with canister, and fire into Humphreys' retreating regiments.

A second Federal brigade, trying to turn the Confederate left flank, also is driven back, leaving its dead and wounded within yards of Cumberland Church.

The 5th New Hampshire Infantry Regiment runs headlong into the Confederate counterattack. Lashed from three sides and low on ammunition, they're forced to fall back, leaving behind their regiment's flags along with more than fifty soldiers and five officers, including twice-wounded Captain John S. Ricker.

In the smoke and confusion of the Confederate counterattack, Private Henry T. Bahnson, Company B, 1st North Carolina Infantry Regiment, his weapon empty and not a round in his bullet pouch, has lost his squad.

Alone, frightened, and having decided that he's done all he can, he's resigned himself to surrendering to the first Yankees he meets.

Coming upon a railroad cut, as his bayoneted rifle appears at its edge, Bahnson is surprised by the shouts of Union soldiers in the cut.

"Don't shoot, Johnny! Don't shoot! We surrender!"

Peering over the edge Bahnson sees dozens of Northern soldiers at the base of the cut, most of them already throwing down their weapons and raising their arms.

He gulps then levels his empty rifle in their general direction and, in as stern a voice as he can muster, calls back.

"All right. The rest of you Yanks throw down your rifles. Quick now. Then make a single line and come on out of there. Quick now or we'll shoot."

Then, encouraged to see that they're actually doing it, he continues to threaten them with his empty rifle as he prods, "Hurry up now. Step lively."

His prisoners march from the cut, and he's busy forming them into a double line when he's joined by another Confederate private.

"What you got there, Henry?" the soldier calls.

"Got me a whole passle of Bluebellies," Bahnson grins.

"Well," his friend grins back, "reckon we do too," and he points to the rest of their squad, now coming up with more than a company of captured Union soldiers.

One of the Yankee prisoners scratches his head in surprise, "Why is this all of you? Ten of you fellers? You yelled so much back there that we thought the whole Reb army was comin' at us!"

Safely behind their lines the handful of Confederates count their prisoners: twenty-one officers, a dozen sergeants, and 103 infantrymen of the 59th New York and the 7th Michigan Infantry Regiments.

Meanwhile, below Cumberland Church, Fitz Lee's cavalry, supported by Confederate infantry that includes the famed Texas Brigade, 1,500 strong at Gettysburg but down now to 130 men, have caught Federal General J. Irvin Gregg's cavalry brigade strung out above an Appomattox ford west of Farmville.

Gregg is caught by surprise. His riders, restricted by the narrow, fence-bordered road into a long column, are suddenly struck by Confederate rifle and canister fire.

The canister, fired at close range down the narrow road, is devastating, tearing men and horses apart.

Gregg's following riders can't clear away the high-rail fences to escape to the open fields. Worse yet, Fitz Lee's cavalry are sweeping in from those fields to tear at their flanks.

The Union cavalry brigade is cut to pieces and their general taken prisoner.

At one point in the fighting below Cumberland Church, Robert E. Lee is reaching for a battle flag to personally lead the Southern attack when his soldiers seize Traveller's reins.

"No, no, General Lee!" they call. "You go back; we'll do the work." Then, screaming their Rebel yell, they charge forward.

Later, watching the fighting from a knoll, Lee takes time for little things, correcting a young officer who'd ridden across its forward slope to deliver a message.

"Young man, you came up the wrong side of this hill. You must not unnecessarily expose yourself that way."

"General," the young officer is surprised at the rebuke, "I would be ashamed to hide behind the hill when you are so exposed yourself."

Lee's answer leaves no room for more discussion: "It is my *duty* to be here, sir. I *must* see what is happening. Now go back the way I told you."

Humphreys, hearing the heavy firing to the south, believes that it's Wright's VI Corps advancing north from Farmville to strike the Confederates' flank. It's not; it's Gregg's disastrous cavalry fight. Wright, delayed by the destroyed bridges at Farmville, won't be near Humphreys for hours. Humphreys' II Corps is on its own and, as dusk is falling, he calls off his attack. Lee's escape route remains open.

LEE

Confederate artillery Colonel William Poague goes to congratulate General Mahone on their victory.

He finds Mahone under a poplar tree, sheltering himself from a thunder shower. Poague expects Mahone to be in good spirits, but the little general is in a towering rage, cursing Yankee soldiers for all he's worth. From Grant right on down to the lowest private in the Federal army.

A fiery fighter any time, Mahone would like to punish his enemies more this day. Particularly the Bluebellies, he complains to Poague, who'd captured his headquarters' wagon and with it his cow, "Old Charity."

"They took my headquarters' wagon, Colonel; and they took 'Old Charity.' You know her; best cow there ever was. Been in every campaign with me. Never failed me. You know my health is delicate. Can't eat a thing but tea and crackers and Old Charity's milk. And now she's been stolen! It's the worst personal loss I've had in this long war."

"Barbarians, sir! Barbarians!" Mahone continues to storm until Robert E. Lee, who arrives during the tirade, suggests a solution.

"Now, General. Your men were magnificent. They've bought our army time; kept open our march to the west. Don't be disheartened; perhaps the kind Providence that has smiled on us this evening will provide another cow, just down the road. She won't be as fine as Old Charity, of course, but perhaps she'll suffice for now."

He also finds time to congratulate his son, Rooney, on the Gray cavalry's victory. Robert E. Lee is cheerful, smiling, confident, but he quietly draws his son aside to mention, for the first time, the possibility of surrender.

"Keep your command together and in good spirits," he advises the young general. "Don't let them think of surrender. I will get all of you out of this."

Lee's situation is desperate, but he has cause to be encouraged. His men still fight hard; still can drive Federal regiments from the field. He'll not have them think of surrender; he'll not think of it himself. Not yet, not quite yet.

After dark, with the fighting ended, he rests near the church while his soldiers again take up their retreat, west toward Appomattox Station. Gordon will lead; Longstreet, because his troops are fresher, will guard their rear, for the moment at least the point of greatest danger.

GRANT

In Burkeville, Grant does not sleep well. He has much on his mind; couriers come and go, disturbing any rest he might have; and the migraine headache which persisted, dull but steady all day long, has worsened.

Rations are short, he worries; and now he must feed not only his own men and horses but those taken from Lee as well. His engineers are laying track to accommodate the wider Northern locomotives. Once it's linked with his Military Railroad, supplies can be hustled forward.

"Get the cracker line open, Colonel!" he'd ordered and hundreds of men are working around the clock to do it.

The horses are tired, and the men's tempers frayed. Even his generals have been affected. Meade, still confined to his ambulance, is cross as a bear, angrily charging that Sheridan's claiming credit for Sayler's Creek has downplayed Humphreys' and Wright's roles in the battle and, it follows, Meade's own leadership.

As Grant smokes his last cigar of the day, he ponders how he can keep them working together until the job's done.

Then he sighs, "It's enough to give a body a headache," and tells Rawlins for the fourth time that he's going to lie down a bit.

Both Federal armies are up early and moving. Griffin's V Corps will cross from the armies' right to their left, a twenty-five mile march to Prince Edward Court House, southwest of Farmville. There Griffin can block any move by Lee toward Danville.

Ahead of Griffin, Ord's Army of the James will march to support Sheridan. The rest of Meade's Army of the Potomac, Wright's and Humphreys' corps, will continue their pursuit above the Appomattox.

Long before sunrise, Grant gives up on his own rest and rides Cincinnati toward Farmville.

Soon he's passing regiment after regiment marching hard through an early morning rain. The road is ankle deep in clinging red gumbo, and Grant rides slowly between the columns so he'll splash no more than necessary.

As he passes, soldiers study the dark-bearded man on the large black horse. He's wearing a mud-spattered slicker over his mud-spattered private's uniform. There is no sign of his rank. His face and beard are caked with fresh-dried mud; a dead cigar, dripping water, is clenched in his teeth. From head to toe he looks as if he'd not slept for several nights and then probably rolled up in a saddle blanket beside the road.

Yet somehow it also seems to them as if he's right where he's meant to be, riding in the rain between long columns of men and knowing exactly where he's going and what he means to do when he gets there.

When a soldier recognizes him the word spreads: "Grant! That's Grant! The Old Man's headin' up front. That's why we're hurryin' so. Looks like he knows what it's all about, now don't he!"

There isn't much cheering as Cincinnati passes. Too cold and wet for that; and Grant's not the kind of man who encourages cheering. McClellan now, the veterans recall, he's the kind of man that you cheer. But this fellow Grant, "Well, trappins' like that don't seem to make no never mind to him."

But, if they don't cheer, they do grin at him and tease a bit.

"Sheridan's cavalry's givin' out, General. We'll catch up with 'em over the next hill."

"Infantry'll crush this mud's far as you want us to go."

"We've marched twenty miles on this stretch, General; reckon we're good for twenty more if you say so."

"We're goin' by shank's mare, General, but we'll get there all the same."

Ignoring the rain and the worsening headache, Grant raises his hat or offers a shy smile or a nod to eyes he catches; the sort of thing one man does with another. All the while Cincinnati continues moving along, up the line. Up, somewhere beyond that next range of hills, where all morning they hear the rumble of distant artillery.

A little before noon he reaches Farmville and sets up his headquarters at the Prince Edward Hotel, a comfortable brick building just south of the Appomattox.

Then, having done all that he can for now, he takes a chair on the hotel's broad piazza where his cigar smoking won't bother anyone. He'll sit there most of the afternoon, attending to messengers and listening to the sound of fighting somewhere north of the river: Humphreys' infantry, above the town; and Crook's cavalry, Gregg's brigade leading, a couple miles northwest of the town, are fighting hard.

"If Humphreys can get there while Crook's coming in from the south," he shares with Rawlins, "we may we able to end this fight right here."

The dull thump of cannon and the sharp crackle of rifle fire grow louder, however, and soon both Humphreys and Crook have sent for help.

"Not yet, not yet," Grant shakes his head. "They're doing all right. Keeping Lee busy while Sheridan gets ahead of him once and for all."

By late afternoon, however, when it's apparent that both Crook and Humphreys have been repulsed and that Lee is counterattacking, Grant turns to Wright.

"General, see if you can't get your artillery across that ford and throw some planks across that railroad bridge for your infantry. Humphreys has got himself a bear by the tail up there. Soon's you can get across that river, drive into their flank."

At dusk, when it's apparent that little more can be done, he orders Humphreys and Wright to pull back.

Lee's lines are still intact; a small victory for the Gray army that alarms Grant's staff officers.

Grant, however, is his usual pleasant, maddeningly calm, uncommunicative self. He asks no questions, solicits no advice; just sits on the brick piazza, legs crossed, hands in his lap, smoking his cigar and whittling; thinking.

Rawlins calms the others. "Never mind. If he wants you to do something, he'll tell you. I expect he's considering that map he's carved in his mind. When he makes up his mind, he'll let us know. Until then, you couldn't budge the man with a mule. Just leave him be."

He's right. Grant, puffing his cigar (he's emptying his case early today, Rawlins frets), is seeing in its smoke his units spread out across the Virginia countryside. Seeing them and the country they travel as clearly as if he were there.

Griffin's corps will be at Prince Edward Court House now. Better than twenty-five miles today and that after more than thirty miles yesterday. Not bad at all. And Griffin won't rest more than a couple hours before he pushes after Ord and Sheridan. He's still got his hackles up about Sheridan's asking for Wright's corps instead of his. Means to show him who's best; and I expect he will.

Sheridan: Major Young's scouts tapped into a telegraph line clicking about trains coming from Lynchburg with rations. They'll draw Lee to Appomattox Station. If Sheridan can get there first, he can take the trains and block the Lynchburg road. No point in losing any more men here; let Lee's men wear themselves out with another hard march while Meade pesters their rear and flank some more. Then, maybe, when they see that Sheridan's in front of them, we can end this without another big fight.

Ord and Griffin have suggested that he send a message calling on Lee to surrender before more lives are lost. Grant had nodded at the idea. Earlier, a captured Confederate surgeon, a friend of Confederate General Richard Ewell, had ventured the same idea. He's thinking on it.

At dusk he sits quietly on the hotel's piazza, its lanterns outlining his stubby features, watching Wright's men as they plank the railroad bridge.

A soldier recognizes the lone figure and shouts.

"Grant. Grant. It's the Old Man," and the word spreads like wildfire.

This time the passing soldiers cheer him and, when he hears their cheers, the shy, complex man, so many times a failure in his personal life, instinctively stands, doffs his hat, and moves a few steps closer to them.

When he does, entire regiments, recognizing this undemonstrative general who'd always seemed untouched by the trappings most other generals favored, feel something that they've never quite felt before. In that moment they truly are *his* army, and he belongs to *them*.

Their shouts echo back along the column of tired men, suddenly not so tired after all. Regiment after regiment take it up, spreading the word.

"Grant! Grant's up ahead by that hotel. He's watchin' us. Needs us; that's why we're marchin' north. Grant himself."

A regimental band hurries up, then another, and another. As the passing regiments return Grant's waves with cheers and waves of their own, bandmasters call for "John Brown's Body." Then the "Battle Hymn," and a dozen more of the songs they've adopted over the past four years. Soon an entire infantry division is singing, "We'll hang Jeff Davis to a sour apple tree."

Men dart from the ranks to improvise torches from the bonfires lighting the street. Then they wave the torches as they march and continue cheering. The solid, deep, rough-cut cheers of veterans who've seen the elephant too many times to feel that they *have* to cheer anyone or anything. Men no longer fooled by commanders who court their cheers in their camps then, by their incompetence, bleed them white on battlefields.

They've taken the measure of this man Grant, however, and like what they see. And now it's pretty clear that he's taken the measure of them too. Something Burnside, or McClellan, or Hooker had never done.

Their night march becomes an impromptu, grand review, with Grant as the reviewing officer. Something no one planned but which none of them will ever forget.

When Wright's men have passed and the night becomes quiet again, Grant takes a final reflective puff on his cigar and turns to go inside. There's more light there and, as he passes, he shares with Rawlins, "I've a mind to suggest to Lee that he surrender. My asking can do no harm."

A little later he calls for his Adjutant General, Brigadier General Seth Williams.

"General, get this to General Lee. You'd best use Humphreys' front as he's closest to the Rebel rear. Have it sent into their lines. I'll be here tonight, should he reply."

That done he turns to the hotel clerk and, as if he were an ordinary traveler spending a single night in the little Virginia community, requests a room.

"Yes, sir. It's all been arranged by General Rawlins. You shall sleep in Number 4, the same room General Lee himself occupied last night."

That's a lie, but it doesn't seem to impress Grant one way or the other anyway. He merely nods his thanks, takes the key he really doesn't need, and plods wearily up the stairs.

Williams is off on a long ride, four or five miles downstream to the wagon track road below High Bridge then west to Humphreys' headquarters. Then II Corps' guides lead him to the Confederate lines below Cumberland Church.

About nine o'clock, Confederate Colonel Herman H. Perry is called to investigate a reported flag of truce before Mahone's lines. Confederate infantrymen there are skittish and, not an hour before, they'd fired on several Federal riders approaching their lines. One of the riders had fallen and, in the darkness, they'd heard someone curse and call out, "Don't you know what a flag of truce is? You've fired on one and killed one of my men."

A Confederate sergeant calmly called back, "Didn't see no flag of truce, mister. Gotta see one to respect it."

Then he'd sent for Colonel Perry. Whatever's out there, let him handle it.

Perry buckles on his sword and revolver then cautiously steps beyond Mahone's breastworks. He walks forward slowly, carefully. Yankee soldiers are out there somewhere, and the ground still is dotted with the bodies of men who'd fallen that afternoon.

Then he sees several horses, their riders dismounted, waiting.

"Flag of truce," he calls, "over here!"

From the woodline a dark figure shouts in return. Then a Federal officer approaches.

The officer, a general no less, speaks first, "Colonel, I'm General Seth Williams of Grant's staff."

Perry, acutely aware of his dirty, threadbare Confederate uniform and slouched hat, so poor in comparison with Williams' appearance, draws himself as tall as he can and assumes the expression of a man very well satisfied with his appearance.

"Colonel Perry, General. How may I help you?"

Williams, taking Perry's measure in the dim light then smiling in a friendly way, suggests, "Colonel, may I offer you some very fine brandy from my flask?"

Perry, tired, disheartened, hasn't tasted solid food for two days and has no prospect for any except the several handsful of corn he'd stuffed in his coat pocket for when he can find time to roast them. He'd like that drink in the worst way, the worst way. Had another Confederate offered it, he'd have accepted it in a second. But no, he decides, I'll not take it from a Yank.

Stretching his lanky frame an inch or so higher than his alloted six feet, he bows and politely explains, "Sir, I'm authorized to hear you out, but I can't properly accept or offer any courtesies beyond that."

Perry, hoping to give the impression that he's just dined on caviar and champagne and really can't handle a bit more, doubts that he's succeeded. The Yankee general, however, smiles, nods his understanding, and returns the flask to his pocket. Perry sighs, knowing that, if Williams had opened the flask to take a sip himself, he'd have smelled the brandy and surrendered without a fight.

"I have this letter," Williams hands it to Perry, "from General Grant to General Lee. Please get it to him as quickly as possible. Should the General wish to reply, I'll wait here for his answer."

"As you wish," Perry takes the letter. "Is there anything more?"

"No, I believe not."

The two officers formally bow, and Perry has turned toward Mahone's lines when Williams calls him back. A messenger from Humphreys' headquarters has come with personal letters and photographs taken from General Mahone's captured wagon that afternoon. "Old Charity," Mahone's cow and the object of Mahone's continuing ill-temper this evening, apparently will not be returned.

Later, Lee studies Grant's message:

Headquarters, Armies of the United States
April 7, 1865 - 5 p. m.
General R. E. Lee,
Commanding C. S. Army:
General: The results of the last week must convince you of the hopelessness of further resistance on the part of the Army of Northern Virginia in this struggle. I feel that it is so, and regard it as my duty to shift from myself the responsibility of any further effusion of blood, by asking of you the surrender of that portion of the C.S. Army known as the Army of Northern Virginia.
Very respectfully, your obedient servant.
U. S. Grant
Lieutenant-General
Commanding Armies of the United States

When Lee finishes reading the letter, he hands it to Longstreet. Longstreet reads it twice then simply shakes his head and mutters, "Not yet, General; not yet."

Lee nods and turns to his field desk to scratch out a reply. He doesn't share his answer with Longstreet, but when he makes no change to their marching orders Longstreet knows that Lee's rejected Grant's offer.

By 10 p.m. Union General Seth Williams, waiting in the moonlight between the Confederate and Federal lines, is galloping back to Grant's headquarters.

Well after midnight, Rawlins brings Lee's reply to Grant's room. Grant, plagued by his headache, is lying down but is not asleep.

"Lee's answer?"

"Yes," and he hands it to Grant who uses a lantern to read the message:

7th Apl '65

Genl

    I recd your note of this date. Though not entertaining the opinion you express of the hopelessness of further resistance on the part of the Army of N. Va. I reciprocate your desire to avoid useless effusion of blood & therefore before considering your proposition, ask the terms you will offer on condition of surrender.

Very resp your obt. Servt

R. E. Lee

Genl

When he's finished reading, Grant hands the message to Rawlins, without comment.

Reading it, Rawlins, however, angrily snorts, "Stalling. He's stalling, Grant. He won't surrender until he's made to surrender."

"Never mind," Grant waves a weary hand. "It's not as I hoped, but he's left the door open. When my head can better handle it, I'll answer."

SHERIDAN

In the morning Sheridan sends Custer's and Devin's troopers on a muddy ride west to Prince Edward Court House while Crook fords the Appomattox to strike at Lee's flank.

Custer, with a couple hours' sleep, is in high spirits as he talks with Confederate General Joseph Kershaw and Colonel Frank Huger. They've shared Custer's blankets and a quick breakfast that Eliza, Custer's personal cook, served from the back of Custer's headquarters' wagon.

Kershaw and Huger, with the other captured Confederate officers waiting to be moved to City Point, are saying good-bye to Custer when the flamboyant general's escort arrives: a score of riders, each carrying one or more Confederate battle flags. Thirty-one captured banners in all. Custer explains that each day soldiers who had distinguished themselves are selected by their commanders to be his personal escort. This morning they carry battle flags taken at Sayler's Creek.

Kershaw, seeing the flags of some of the most legendary units of the Confederate army, now carried by their enemy, is stunned and saddened, as if he's watching a terrible funeral procession.

As he turns to ride from his camp, Custer, sensing Kershaw's reaction, nods, and his headquarters' band, all mounted on gray horses, strike up "The Bonnie Blue Flag," one of their prisoner-guests' favorite marching songs. Custer adds his personal salute with a wave of his hat, and several of his troopers drop hardtack-filled haversacks at Kershaw's and Huger's feet.

Enemy or not, Kershaw decides, he's a gallant man. Waving his hat, he shouts, "There goes a chivalrous fellow, boys! Let's give him three cheers!"

Rebel yells echo in the morning's quiet. Then, as Custer's caravan rides west, his band plays a final tribute, the lilting, blood-stirring notes of "Dixie" slowly fading into the mist.

During the day Sheridan has a personal encounter with an unreconstructed Confederate, an ancient civilian whom Sheridan finds calmly rocking back and forth on a porch. The Southern patriot, with long gray hair tumbling to his collar and wearing a Prince Albert coat, brown linen vest, buff cotton trousers and red morocco slippers, eyes Sheridan but says nothing.

Sheridan returns the look then asks, "Any of Lee's troops passed here?"

"Sir," the elderly gentleman sitting ramrod straight in his rocker, has anticipated the question, "I've seen none. But if I had, I'd refuse to answer. I will not give you any information which might harm General Lee."

"All right," Sheridan grunts, "let's try something else. How far is it to the Buffalo River?"

"I don't know."

"The devil you don't," Sheridan's dark face turns a deeper red and his voice rises. "How long have you lived here?"

"All my life," the defiant answer.

"Very well, sir. It's time that you found out just how far it is."

Calling to a nearby officer, he orders, "Captain, walk this gentleman down to the Buffalo River. Show it to him; then walk him back."

By sundown Sheridan's located something more important to him than the river. Major Young's scouts have learned that several trains with rations for Lee's army have left Lynchburg for Appomattox Station.

He relays the information to Grant and orders Custer to force-march to Appomattox. The rest of Sheridan's cavalry will follow. He also sends word to Griffin and Ord: don't camp; push on. Get to Appomattox as quickly as possible.

When Grant's courier arrives, approving Sheridan's striking for Appomattox and informing him that Grant has sent a note to Lee suggesting surrender, Sheridan scoffs.

"He can try, Colonel Newhall, but Lee won't surrender until he has no other choice. None whatsoever." Then he orders, "Send riders out. Push everyone along. I want that Lynchburg road blocked at Appomattox Court House tomorrow."

LINCOLN

During the morning, as he awaits the *River Queen* returning Mary Lincoln's party from their visit to Richmond, Abraham Lincoln studies the night's dispatches. They continue to be optimistic, and he delights in relaying to Stanton news of the victory at Sayler's Creek, ticking off the names of captured Confederate generals and telling Stanton that, following Sheridan's lead, he's directed Grant to "let the thing be pressed."

He's in good spirits, despite two troubling incidents.

Vice President Andrew Johnson had appeared at City Point on an army packet boat, sending word that he wished to call upon the President before making his own tour of Richmond. Lincoln, having read a recent speech by Johnson calling for vengeance against the South, doesn't want to see him.

"It would serve no purpose, Porter," he explained. "We're too far apart on this question. We need reconciliation with the South, not vengeance. Richmond or Washington, Mr. Johnson can get along without me. Tell him that I am busy with other matters."

"I fear," he added, "that we are seeing the first of a stream of visitors anxious to crow over Richmond. Too bad."

And he was troubled by the sight of another boatload of Confederate prisoners: ragged, dirty, sharing scraps of bread they've taken from their haversacks.

"Poor men, poor men," he shook his head. "But at least they've survived and now can rebuild their lives."

This morning, however, the good news from Grant has pushed all that aside. He meets Mary Lincoln and her guests at City Point's dock; his tall, lean figure towering over them.

During lunch they share impressions of occupied Richmond. The First Lady has not enjoyed their tour. Richmond's streets seem busier than she expected, and its citizens paid little attention to their caravan. A Federal band was playing in Capitol Square, but, other than white children wheeled there by black servants, she'd not seen a single white person enjoying the concert.

And she'd seen many fluttering curtains and the dim outline of feminine features studying her as they passed, but no one parted curtains or came to their porches for a better view. And in the entire visit no one had spoken to her except to answer her questions.

"Not very friendly, Father," she judged; and he patted her arm, "I know, Mother. But we must be patient. It will come in time."

In the afternoon he takes them to Petersburg, taking particular interest in showing them the battlefield he visited while the fighting still raged.
It's obvious that he's enjoyed his brief vacation with Grant and his army
and that he feels that just by being at City Point he's contributed something
to their victory.

He finds that his guests' interest in the battlefield, however, doesn't
match his own. They listen politely but seem not to understand what really
happened there. Not to understand or care; at least not to care as he does.

He stops trying to share his feelings, turning instead to the window as
the others pick up the brighter threads of a conversation his talking about
the battlefield had interrupted.

Beyond the train the Virginia countryside seems from another world,
a world almost completely stripped of homes, barns, fences, and roads.
One broad, barren field, churned again and again by shell fire and by
trenches that seem to run in every direction.

Here and there he sees small, ragged clumps of trees marking where
some headquarters stood a few days before, but overall there are few
trees to soften the harsh landscape. Instead, the country is marked by row
after row of empty soldiers' huts separated by soldiers' roads that seem to
start nowhere and go nowhere. Forts, rifle pits, abatis, chevaux-de-frise,
all deserted but still standing, as if awaiting the return of two phantom
armies. Deserted and as silent as death.

He knows that one day the fields will be green again; the obstructions
taken down and the trenches plowed into fields; the homes and roads re-
built. But it will all take time. Perhaps a generation or more before nature
and man can erase the evidence of what happened here. It would be well,
he thinks, if those who come to crow over Richmond instead could be
shown the trenches below Petersburg. But he knows that won't happen.

That evening he hosts a dinner party aboard the *River Queen*. It's a
beautiful spring evening. The sky is bright with stars, and the James River,
free of the terrible sights and sounds of war it knew just a few days before,
is lit by the lights of hundreds of ships anchored off City Point.

Lincoln enjoys the evening, smiling, more than carrying his share of
the conversation. He entertains them with stories of his prairie days, and of
Washington, and of his experiences at City Point with Grant and his army.
Cheerful, pleasant, enjoying their company; yet, now and then, he allows
his mind to slip away from their casual talk. Admiral David Porter, watching
closely, can tell when that happens for the President will close his eyes and
his features will slip into that look of infinite sadness that they most often
ascribe to him. Then he'll recover, smile at a compliment that the Count de

Chambrun has just paid Mary Lincoln, and only Porter will realize that he's been away.

Overall, a pleasant evening although it ends on an unhappy note when a young officer of the Sanitary Commission, attempting to match the Count, tries to flatter the First Lady.

"Mrs. Lincoln, you should have seen the President on his triumphal entry into Richmond. Every eye was focused on him. Ladies kissed their hands to him and waved handkerchiefs. A glorious moment. He's quite a hero when surrounded by pretty ladies."

Mary Lincoln glares then erupts, "Your familiarity, young man, is offensive to me. Offensive, sir! Very offensive!"

The young officer, flushing to his toes, can't redeem himself. The First Lady unloads all her anger, all her frustrations on him. Then on anyone else who attempts to rescue the situation. Until finally the President pats her hand and quietly tells her, "Now, that's enough, Mother. That's enough."

Lincoln ends the dinner party with the announcement that they will be returning to Washington the next evening. His brief vacation is over; there are important matters he must attend to in the capital.

THEY FOUGHT FOR THE UNION

President Abraham Lincoln

**Lieutenant General
Ulysses S. Grant, USA**

**Major General
William T.
Sherman, USA**

**Major General
George G.
Meade, USA**

Major General
Edward O. C. Ord,
USA

Major General Philip
H. Sheridan, USA

Major General George
A. Custer, USA

Major General
Joshua L.
Chamberlain, USA

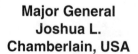

Rear Admiral David D. Porter, USN

Major General Charles Griffin, USA

Major General Philip H. Sheridan, USA, with Brigadier General James W. Forsyth, Major General Wesley Merritt, Brigadier General Thomas C. Devin, and Major General George A. Custer

Lieutenant General Ulysses S. Grant, Headquarters, City Point, Virginia, 1864

**President Abraham
Lincoln with wife, Mary,
and sons (captain)
Robert and Tad**

THEY FOUGHT FOR THE CONFEDERACY

President Jefferson Davis

General Robert E. Lee, CSA

General Joseph E. Johnston, CSA

Lieutenant General James Longstreet, CSA

**Lieutenant General
Richard S. Ewell, CSA**

**Major General
Fitzhugh Lee, CSA**

Major General John B. Gordon, CSA

Major General George E. Pickett, CSA

General Robert E. Lee with his son, Brigadier General Washington Custis Lee, and aide, Lieutenant Colonel Walter H. Taylor, Richmond, 1865

The McLean House, Appomattox Court House, Virginia

The High Bridge, Farmville, Virginia

**The surrender of the Army of Northern Virginia,
April 12, 1865; painting by Ken Riley**

CHAPTER 6

He Who Marches Fastest   Saturday, April 8, 1865

THE PURSUIT

Grant's up early; restless, and the migraine headache a persistent threat. He's sent General Williams off with his second message to Lee and is reviewing the orders he's given to press Lee harder.

Below the Appomattox, Sheridan is pushing west with Crook's, Custer's, Devin's, and Mackenzie's cavalry divisions. Following Sheridan are Griffin's V Corps and Ord's Army of the James. Above the river, Humphreys' II Corps and Wright's VI Corps pursue Lee's retreating army.

As they ride to join Meade, Rawlins continues to lecture Grant about Lee's reply to Grant's first message.

"He's stalling, Grant. Means to slow us down while he hurries off. Maybe get a better deal for his officers when he must surrender. But I'll bet my boots that he won't surrender until he's *made* to surrender. You must press him."

Grant nods but doesn't comment. Already his headache is worsening.

It had been a cold, frosty night with light flakes of snow. As Grant's party nears Cumberland Church, he sees the near-naked bodies of Northern soldiers, killed in yesterday's fighting. Needy Confederates have taken their coats, their trousers, their boots. The soldiers' pale skins seem very white against the reddish-brown ground before the church.

"Rawlins," Grant mutters around his dead cigar, "see that Wright gets this cleaned up soon's possible. Let's push ahead and find Meade."

By mid-morning, however, the day seems brighter to them. The sun has warmed the air; the clouds have opened to a deep blue spring sky; and to the west they see a beautiful land opening before them. A land that already seems untouched by the war. Roads that are not so worn; and more farms, many of which, unlike those behind them, are freshly planted.

As Grant rides, couriers arrive, assuring him that all his units are marching against scattered Confederate resistance. General Seth Williams, returning from delivering Grant's second message to Lee, reports, "Had a little trouble, General; the Rebs are pretty skittish. Twice they fired on my flag of truce before I could get them to recognize it."

He nods, appreciating Williams' problem, and rides on. Despite the headache, his spirits rise. So far the day is going well.

An hour's ride and they come upon Major General Andrew A. Humphreys, a spare man with iron-gray hair and moustache, bluish-gray eyes, and habitually wearing a kindly expression. Humphreys, from the Old Army, is a disciplinarian and a hard fighter, but he's well liked by his men. Seldom excited, except when he finds that an order he's given hasn't been carried out. Then he's known for flaming outbursts that cause any lost time to be made up pretty fast.

Grant finds Humphreys near a crossroads west of Cumberland Church, down on all fours closely studying the hoof-torn road a few inches below his wire-rimmed spectacles, much like a hunting dog seeking a scent.

"Well, General," Grant chuckles, "what do you think?"

"I think, sir, that the main body went that way," he points. "And I shall follow them. Let General Wright go after the others."

Grant nods approval and watches Humphreys ride forward.

The countryside, slowly rising to the west, is changing. Before them, better than forty miles away, they see the dark, brooding shoulders of the Blue Ridge Mountains; behind them is the much flatter watershed that drains into the James and eventually into the ocean. Wheat already is nearly a foot high here; peach trees bloom; and cherry and plum trees are white with blossoms.

They find Meade's staff clustered around the canvas-covered wagon bearing their general. Meade, still ill, again must ride an army ambulance. Embarrassing for a commander, and Meade's still troubled that Sheridan is claiming credit for their success so far.

As Grant nears, Meade's small, bald head and thick spectacles appear at the opening in the wagon's cover. He waves to Grant and Grant, with unusual warmth, returns the wave and adds, "How are you, Old Fellow?"

Meade reports that Humphreys' corps hasn't yet caught up with Longstreet's rear guard and is pushing ahead as fast as it can. Meanwhile, it's picking up hundreds of Confederate stragglers and wounded, some cannon, and many wagons.

"The situation looks very good, General," he reports. "Very good. We shall follow this fox to his den, sir; to his den."

"That's fine," Grant answers, "but I still hope to cut him off before he can hole up." Then, lest Meade take offense at his reminder, he asks, "And how do you feel this morning? Better I hope."

"No, sir; not better I'm afraid. But I'll keep up. And you?"

Grant, his headache a dull, steady pain, nods over his cigar, "Toler'ble, General, just toler'ble."

He rides on, fairly close to Meade yet staying apart from him, dispatching and receiving couriers. The headache, however, worsens until by mid-afternoon it's a blinding, nauseating pain that finally gets the best of him.

"Rawlins," he calls, "this headache is terrible. Maybe it'll ease up if I'm not riding. Find us a farmhouse ahead; let's make camp."

Two days before, when Grant rode to Sheridan's camp near Jetersville, they'd left their clothing, bedrolls, and camp equipment behind. They've not returned for them, eating with whatever unit they're near and camping wherever they can find shelter. So far as Grant is concerned, one place is as good as another.

Rawlins selects the Crutes' farmhouse, called "Clifton," near Curdsville for Grant's headquarters. Couriers can find it easily; and it's less than a mile from Meade's headquarters for the evening.

Grant hasn't heard from his second message to Lee but quietly dismisses Rawlins' alarm, "Now, Rawlins, don't sour your milk over it. His army's all strung out. Its head's somewhere near Appomattox, and I expect that its tail's God knows where. They may have trouble finding him. We'll give him a bit longer."

Humphreys reports that he's picking up more and more Confederate stragglers, soldiers too tired to continue the march; wounded men left beside the road; and deserters waiting to be taken out of the war. Rifles, with their bayonets driven into the ground (the usual sign of surrender), are appearing more often. Usually it's a lone rifle or two but sometimes, where entire squads have given up the fight, they're grouped like small, stripped trees. Artillery crews are surrendering with intact cannon and limbers; and hundreds of wagons and ambulances, crammed with wounded of both sides, have been abandoned, their teams no longer able to pull them.

Confederate General John Gordon sends four of his surgeons under a flag of truce to help Humphreys' surgeons caring for Gordon's men left beside the road. Humphreys thanks the Gray doctors but, having plenty of surgeons of his own, sends them back to Gordon who undoubtedly needs them more.

———

Yankee riflemen find more and more graves beside the road but, suspicious of them, check a patch of freshly-turned earth near a broken artillery caisson. They've not dug far when they uncover a 12-pounder cannon. In nearby "graves" they find three more.

———

Because this part of Virginia has not seen the fighting that's ravaged its northern and eastern counties, there's a much better chance for soldiers, Blue or Gray, foraging for food. Local farmers, however, know about hungry soldiers and try to outwit them.

W. D. Jones, an octogenarian not about to part with a keg of his favorite brandy, buries it under a suitable tombstone in the family plot on his farm. Humphreys' men don't find the brandy, but they do find the hams, and pork, and bacon Jones also hid in tree branches.

———

Because Lee's soldiers aren't apt to leave anything worthwhile for Yankee infantrymen, Humphreys' men are surprised to find a large, carelessly hidden barrel marked "Sorghum Molasses" near their march.

Word of the priceless treasure quickly spreads, and the best part of a Yankee rifle company quickly throng around the barrel. Then the lucky ones in front dip their tin cups or canteens into the dark, appetizing liquid before hurrying off to find a quiet place by a fence rail to enjoy the unexpected feast.

They savor their first, deep drink but, when its taste comes through, spit out what's left.

"Don't rush to drink it, boys," one laughs. "Tain't molasses. It's cold tar. Laughs on us. Johnny's fetched us up short this time."

Some cuss about it, but most of them enjoy the joke so much that they decide to share it, moving the barrel to where it can be easily seen from the road. Then they hide and watch as following Yankee regiments find the tar barrel, also bless their good fortune, then finally realize the joke that Johnny Reb's played on them. The "treasure" will find its way through the best part of a Northern division before the trick plays out.

———

Late in the afternoon, Humphreys' riflemen catch up with Longstreet's rear guard. Intermittent skirmishing continues until the light fails.

At dusk Humphreys' corps, having marched more than twenty-five miles, bivouac within a dozen miles of Appomattox Court House. They're to take up the march again at 4 a.m.; Humphreys wants to reach the Court House by daybreak.

His soldiers, exhausted, pretty much drop where they are. Somewhere up the road they hear the dull thump of artillery and know what that means. Someone's fighting up there. Maybe Sheridan's gotten ahead of Johnny.

"Whatever it is," the veterans advise, "get yourselves some rest while you can. Reckon we'll be fightin' tomorrow."

CHAMBERLAIN

A dozen miles below Humphreys' men, Sheridan's cavalry and Ord's and Griffin's infantry will march hard all day.

For General Joshua Chamberlain the march begins well enough. Sergeant McDermott wakens him at 4 o'clock with his usual cheerful long burst of fire: "Good mornin', General, darlin'; mornin'; and here's some hot coffee, sir. Colonel Spear's up and about. Cold, sir, but the day's promisin'."

Then, as Chamberlain begins to feel the hot tin cup burning his numbed fingers and the scalding hot coffee coursing through his stiff, aching body, the faithful sergeant continues the argument he'd begun the day before.

"Now, sir, the General insisted on walkin' most of the way yesterday, and I 'spect the General, if he was to tell the truth, has to be feeling every mile of it this morning. Especially with that wound in his hips still not healed and that thump in the chest he got back there near White Oak Road. A close call, sir, a reminder not to tempt fate. And, if there's no other reason comes to the General's mind, may I suggest that the damned horse is gonna feel neglected if he's not rode a bit today? Would the General listen, just this once, sir?"

Then, having done his best, he'd readied Charlemagne and reported to Colonel Spears and the others. The staff, as Chamberlain well knows, will be waiting for Sergeant Mac's report because, if the General walks, they must walk too. And the day promises to be a long one.

By 5 a.m. they're on their way. An easy, steady march with the sun slowly warming their backs, and the sandy road holding fairly firm for easier marching. There's only Sheridan's cavalry ahead of them so the road's not so badly torn for fast marching.

In the distance they see the shadowed ridges and deep crevices of the Blue Ridge range and, a mile or so to their right, the Appomattox. Beyond the river, they catch occasional glimpses of Confederate infantry paralleling their own march.

Once in a while they see Confederate stragglers in distant fields, avoiding their flank guards. If Johnny's not doing any harm, however, they ignore him.

"He's not carryin' a rifle no more," they reason. "Goin' home. Let him be."

At Buffalo River Chamberlain stops beside the bridge to watch his regiments pass. As he waits, he allows Charlemagne to nose knee deep into the creek to drink.

Thoughtfully watching his infantry crossing the narrow bridge above him, Chamberlain doesn't realize that the stallion is wandering deeper into the stream until the horse steps into a deep pool. Horse and rider disappear beneath the cold, dark water. Then they surface, sputtering, near the center of the creek.

As Chamberlain's staff rush to the bank, not sure how to help horse or rider, Charlemagne, confused and frightened, can't find any footing. Finally Chamberlain, swimming with one hand while holding the reins with the other, manages to turn the horse around. Then half-pulling, half-carrying the big horse, General Joshua Lawrence Chamberlain, covered from head to foot with water and thick, black mud, stamps ashore.

On the bank his staff, trying hard not to laugh, wait for his reaction. Except for Sergeant Thomas McDermott.

"General Chamberlain, sir," he salutes and formally addresses the dripping general. "If the General would like to step over to the fire there and get out of those wet clothes, we'll see if we can keep him from getting pneumonia. And now that I've seen the trouble the General can get into riding that damned horse, he'll hear nothing from me if he chooses to lead him all the way back to Portland."

Chamberlain stands beside the fire, draped in a blanket while his uniform dries and the mud is washed from Charlemagne. Meanwhile, each grinning regiment does an "Eyes Right" as they pass their near-naked General. Not as big a parade as Grant's in Farmville, nor as spectacular, but one they'll recall for a long time.

Chamberlain's men almost enjoy the early part of their march. The country, untouched by the war, is blossoming with spring. Foraging is better, even though Sheridan's cavalry ahead of them seem to have cleaned out all the chicken coops in sight. Tobacco, molasses, bacon, cattle, sheep, pigs, poultry, meal, however, still can be found by enterprising soldiers.

At Prospect Station they come upon the Confederate train which had fled Farmville before it could unload all its rations for Lee's army. It escaped

Yankee infantry there, but Crook's cavalry caught it west of the town. Then they stripped it of all the food and clothing they could use. And when Crook's cavalry finished, Ord's infantrymen marched in to take up where the cavalrymen left off. By the time Griffin's corps arrives, there's not much left for them.

They're not too happy about that; unhappier still when Ord's men and wagons cut in between them and Sheridan's cavalry.

It doesn't make much sense to them: they should be up near Sheridan, not Ord's men. Why they've seen Ord's regiments march before, and they all agree that, "Those fellows are God-awful slow."

That's reason enough to spoil their day, but there's more. Ord's supply train also cuts in ahead of them, slowing Griffin's men to a crawl and tearing up the road they must march. So begins a twelve-hour, stop-and-go, hurry-up-and-wait march they'll always remember.

Late in the afternoon word spreads that Major Young's scouts have found more trains with Confederate rations at some place called Appomattox Station, up the road quite a piece. Johnny's headin' that way, the rumor goes; needs those rations bad. Well, they decide, so do we. And, if we can get there first, we'll take those trains. Then we'll block the road, and Johnny'll have to stand and fight or give it up.

Their commanders halt long enough to deliver little speeches about how they have to march harder than ever before; speeches that really aren't necessary because the men have figured it out for themselves: if we get there first, we're going to end this thing. Once and for all. This road, this road we're on now, is the one we've been looking for all this time. This is the road to end this damned war. This is the road home. Simple enough. Their pace quickens.

When General Ord stops to speak with little groups of men, man to man, they listen. Riflemen have to listen to generals; it goes with the job. Besides, they like Ord. He seems to understand more than most officers what it's like to carry maybe thirty pounds of equipment a heap of miles.

Ord understands other things too. As when the 11th Maine Infantry Regiment drops in its tracks for its hourly break, and one of Ord's staff officers shouts, "You men, give way; get out of the road so the General can pass."

Ord straightens the captain out pretty fast: "Stop that, sir! The men are tired. Rein your horse to the side of the road. Go around them."

"That Ord's all right," they agree. "Lucky to have him, 'stead of some of the other generals." And all day Ord rides back and forth along their column: teasing, cajoling, ordering, reminding.

"Legs will win this battle, men!"; "The campaign's in your legs, boys!"; "He who marches fastest will win, remember that!"; "One good, steady march, men, and this war is over!"

Ord's and Griffin's men march all afternoon, then all evening, then into the night. A hard enough march for the lucky ones at the head of the long column but very hard for those trudging in the rear: hurrying fast for a quarter mile then forced to stop while Ord's long line coils and uncoils. Also forced to wait while soldiers pull mired wagons from soft spots in the road. Then they must hurry to catch up. Every hour bugles sound, the column halts, and the men have a few minutes' rest before the bugles sound again. Hardly enough time to ease one's rifle and pack from his shoulders, let alone enough time to stretch out or to sit down comfortably. Just enough time to be frustrating.

It's worse during the night. They can't see where they're going, and Chamberlain's men, behind Ord's long wagon train, have a particularly hard and frustrating time. Tempers are frayed, and infantrymen argue with teamsters about rights-of-way.

At the head of the V Corps' column Chamberlain's infantrymen begin to help Ord's teamsters along, prodding horses and pack mules with their rifle butts and bayonets. An artillery battery, hurrying to catch up with the rest of its battalion somewhere ahead, tries to force its way between Chamberlain's marching men. Bayonets again prick horse flesh, horses scream, and teamsters retaliate with their whips. In a darkly wooded defile, with no room for anyone to step aside, tempers and patience in the 20th Maine Infantry Regiment suddenly snap and a fist fight develops. Another incident they'll remember when all this has passed.

Officers try, with little success, to break up the fight. Each side is busy taking out all its frustrations on the other. A staff officer has several regimental bands hustle to the edge of the melee and strike up a tune to calm the brawling mob of men. It might have worked, but their choice of tunes isn't the best for when the struggling men hear the lilting strains of "The Girl I Left Behind Me" they only fight harder.

It goes on until most of their frustrations have been worked out and everyone is too tired to continue. Then the brawl simply flickers out.

They take up the march again only to interrupt it a short way up the road with another angry exchange between teamsters and infantrymen.

When Ord finally halts his divisions for a short bivouac, Griffin's V Corps passes through Ord's regiments to take the lead, just to show that they can do it.

Behind them Ord's men settle for a few hours' rest. In one of the regiments an orderly spreads a blanket for Captain Daniel Barnard. A few minutes later, however, he finds that another man has wrapped himself in it.

The angry orderly pulls the trespasser's leg and orders, "Here you! Get up! You've got the Captain's bed!"

The sleeping man gets up, apologizes, and turns to find another bed. It's Ord, and he and his corps are about marched out: eighteen miles on the fifth, twenty-three more on the sixth, another twenty-seven yesterday, and more than thirty miles today.

Meanwhile, Griffin continues another five miles until, having made his point, he too calls a halt. To a man and to a horse, they're exhausted. They've marched better than thirty-five miles and feel every one of them.

Their head is five miles below Appomattox Station, but stragglers may be another five or six miles back. The 198th Pennsylvania has lost about half its men; and the 20th Maine has only seventy-five men left. They'll all catch up later.

Riflemen drop their packs and sprawl where they are; officers wrap their horses' reins around their wrists and fall asleep beside the road. Their mounts, heads so low that their noses almost touch their riders' faces, sleep above them.

CUSTER

Not too far ahead of them Sheridan's cavalry have had a hard ride and a hard fight.

They'd begun near Prospect Station, moving west across fields and roads. Sheridan, wearing a double-breasted frock coat and strapped-down trousers outside his boots, and a low-crowned felt hat above his close-cropped black hair, is pushing them hard.

The night before Sergeant James White of Major Young's scouts had come to Sheridan's camp. White, wearing a Confederate uniform, is one of the scouts Young had sent from Jetersville with Lee's message asking Lynchburg officials to send rations for his army. Since then White has been watching for the expected trains.

"I found them a piece west of Appomattox Station, General," he'd gleefully reported to Sheridan. "Sent from Lynchburg after they got my telegraph. Just kinda layin' back, waitin' for some word from Lee that it's all right to come on in. They were 'spicious of me at first, but when I showed them Lee's original order and told them how desperate we were for the rations, they took me at my word. Those train'll be comin' in there tomorrow. But when they don't find the Rebs waitin', they're gonna skedaddle right quick. We gotta hurry up. And we gotta cut that track behind them. Reckon we can do that, General?"

Sheridan, Sergeant White's fellow strategist, reckons that they can and hurries Custer's cavalry toward Appomattox Station.

The warm, near cloudless day and drying road and fields help Custer's march. Beyond a few Confederate stragglers waiting to be picked up, there is no sign of their enemy.

At sunset Custer is watering and resting his horses, about three miles from Appomattox Station, when a scout reports.

"Smoke over the next ridge, General, and train whistles."

Sheridan had ordered Custer to stop for the night, but the aggressive general decides to go for the trains.

He sends three regiments to cut the tracks west of Appomattox Station. Meanwhile, he'll lead the charge on the station itself.

As he gallops toward the station he's delayed when two young women call to him from a large, elegant mansion beside the road, "Help! Help! They're robbing us!"

Custer reins his horse and runs up the walk to encounter a dirty, bearded, burly man in a Federal uniform stepping from the porch, his arms filled with several silver goblets, a silver tray, and a large box. Before the surprised thief can react, Custer's fist stretches him out beside the porch.

Running into the house, Custer sees a second looter trying to escape by a rear door. Grabbing an axe from the kitchen woodpile, he throws it, felling the second man. Then his troopers are all over both men.

As he hurries from the house Custer, a striking figure in his olive green corduroy suit, lavishly adorned with gold braid; the scarlet red, flowing tie with its big "CUSTER" pin; wearing large-roweled Spanish spurs, and armed with a long cavalry saber and heavy revolver, waves his sombrero and calls to the ladies, "I'm sorry; we're not all like that. They'll be punished, I promise. Now I must go."

He turns and gallops toward the heavy gunfire ahead.

Troopers of his 2nd New York Cavalry Regiment are charging the railroad station. Four trains are puffing there but, seeing the Yankee cavalry, their engineers try to back away. One train escapes, rattling back down the track. Custer's men, however, swarm around the remaining three.

A Yankee trooper, Sergeant Fred E. Blodgett of Company K, carbine in hand, swings from his saddle to climb to a locomotive's cab.

"You two, hands up! Right now! One false move and you're dead men!"

Several of Custer's troopers, once railroaders themselves, yank the other trainmen from their cabs. Then, in a bedlam of blowing whistles, ringing bells, and Union soldiers' cheering, they move the captured trains east, away from a possible counterattack.

A grinning Custer glances at his watch. Another hour or so of daylight. Might as well scout the village, he decides, and waves his troopers toward Appomattox Court House several miles east of the station.

The Lynchburg road, which they must hold to block Lee's retreat, will be there.

APPOMATTOX COURT HOUSE

Confederate Brigadier General Lindsay Walker, marching ahead of Lee's army with about a hundred of Lee's reserve artillery pieces, had reached Appomattox Court House a little earlier.

All day he's seen nothing of their enemy; heard no firing. Convinced that he's won the race to the railroad, he has his artillerymen camp about halfway between the village and the railroad station. They're preparing an evening meal when the first Gray cavalry pickets, fleeing the short fight at the station, spread the alarm.

"Yankees! At the station; everywhere!"

A Confederate soldier, galloping his mule as fast as the animal can run and shouting, "The Yankees are coming! The Yankees are coming!" suddenly comes to a deep mud hole. He expects the mule to jump it, and is in mid-shout when the mule decides to abruptly stop instead. The mule slides into the mud, its rider somersaulting through it. Stunned, he staggers from the far side to hide in nearby bushes.

Two of Walker's artillerymen are sharing a hunk of corn bread and some molasses when the fight begins. Frightened horses, many of them riderless, dragging their reins behind them, stampede by their campfire. One of the young soldiers coaxes a mare close enough to spring on her back. The other, seeing a Confederate officer busy shouting orders, steals the officer's horse. Then they gallop west. Their last view of the Confederate army is of the fighting at the station.

Private Jimmy Albright, 12th Virginia Infantry Battalion, has faithfully kept a diary for many months. Seeing that he may be captured and not wanting any Yank to capture his diary, he takes time to tear it into shreds before he too flees.

Meanwhile, Confederate General Walker has reacted quickly. He'd been shaving when the alarm sounded and now hurries about, his face still lathered, shouting orders. Brigadier General Martin Gary's nearby South Carolina cavalry will delay Custer while Walker sets up a barricade across the Court House road.

At the edge of the village he positions his cannon in a hollow square around baggage wagons and ambulances. Artillerymen, now armed with rifles, take positions around the wagons. Then, knowing that whatever has hit the station will be coming his way, he waits.

Major Edward M. Boykin is with Gary's brigade. Earlier that morning Boykin's First Sergeant issued each man a handful of cartridges, the last of their ammunition. Then all day they'd marched, a quiet march in which their enemy seemed simply to have vanished.

Late in the afternoon they set up their camp near the Appomattox railroad station and are there when Walker calls for help.

Gary's little brigade, tired men on tired horses, armed with a few cartridges, charge Custer's regiments. Charge several times, are beaten back, then stubbornly withdraw toward the village. They've bought time, however, for Walker to set up his barricades.

CUSTER

Sheridan has sent couriers to hurry Devin's cavalry forward to reinforce Custer. Custer, however, doesn't mean to wait for them. The enemy is down that road, and he means to attack, with or without Devin.

His red-tied troopers charge down the village's lone street, right into Walker's massed cannon. Repulsed before Walker's makeshift barricade, they form again and attack, again and again.

Custer is everywhere, urging his men to push their attack.

"Colonel Randol," he yells, "go in, go in. Hit 'em hard, and they'll break."

"Give me time to round up all my men, General; then we can do it."

"Never mind the rest of your regiment. Take whatever men you can find and charge those guns. We must take them and the road. Go in, Colonel, go in!"

Again: "Boys, the 3rd Division must have those guns. I'm going to charge if I have to go alone."

And to the 2nd Ohio Cavalry Regiment, "Here's the regiment I want. Come on, you Buckeyes. Take that battery!"

Finally, as dusk settles over the battle area, he makes a final appeal: "Boys, I want a few of you to go up this way with me."

When that attack also is broken, he finally gives it up.

He's a target for a hundred Confederate marksmen. Horses and aides fall around him but Custer, considered lucky by his fellow-generals and by his men, is especially lucky this day. His horse falls, but the flamboyant general's not hit.

During the fight at the edge of the village, a man wearing a nondescript uniform that could belong to either side and riding an equally nondescript mule, rides back and forth between the opposing lines, all the while beating a large kettle drum. No one knows whether he's a Billy Yank or a Johnny Reb; or why the drum, for he shouts nothing to accompany it; or whether he survives. Just one of the odd things that can happen now and then in battle.

Federal horse artillery come up during the fight and their duel with Walker's Confederate cannon becomes one of the closest, fiercest artillery fights of the war. The opposing cannon are seldom more than a hundred yards apart, sometimes down to less than half of that; and their canister shells rip into men and horses. In the gathering darkness cannon flash from muzzle to touchhole every time they fire.

Three times Custer's men charge the Confederate guns; three times they're driven back with many horses and riders down on the village's single street.

Walker's men fire double loads of canister or case shot with fuzes cut to the quick at the mass of dark shadows bearing down on them, knowing by the screams that follow that they've hit their targets.

Private Peter M. Gephard, Company M, 2nd Ohio Cavalry Regiment, shot through the bowels, lies in the street before the courthouse, screaming in agony before he dies.

Despite their losses, however, some Federal cavalrymen reach the Confederate guns to fight hand to hand among them, paying a high price for each one they take.

Lieutenant Colonel Augustus I. Root leads a score of 15th New York Cavalry Regiment troopers in a headlong charge toward the courthouse. They're within eighty yards of it when they run into a solid hail of rifle fire at very close range. Most of the Union riders fall in the first volley. More in the next, and the next, before the survivors give it up.

Root lies among the dead. A sergeant, shot in the stomach, is there too. The wounded sergeant, lying under his fallen horse, begs someone to take his carbine and kill him. That's not necessary, for within five minutes he's dead.

A shell fragment cuts off the first and second fingers of Captain Eric Woodbury's right hand. Then it goes on to tear open his jacket and shirt, finally shattering his left arm.

Frightened, riderless cavalry horses leap the Confederate barricade, wheel into private yards, or flee up the street from where they'd come.

Finally, it's too dark for any more fighting. Heavy smoke and the acrid smell of burnt gunpowder are everywhere. Dead and dying horses and riders litter the street before the Confederate barricade.

The body of a Confederate rifleman, Jesse H. Hutchins, lies near a wagon. His squadmates are stunned at his death. Hutchins had enlisted in the Gray army three days after Southern artillery fired on the Union's Fort Sumter in Charleston harbor, South Carolina, beginning the war.

Hutchins then fought every campaign of the Army of Northern Virginia, serving 1,454 days, only to die in the last twenty-four hours of the life of the army.

A Union bugle sounds "Recall," again and again. Rapidly fading hooves and dimming shouts signal the end of Custer's attack.

Silence, a pause to be sure that the Yankee cavalrymen really have given up the fight. Then an elderly Alabama rifleman steps from behind a tree. He examines Augustus Root's body then, leaning his weapon against the tree, speaks to the rifle as if it were an old companion.

"Well, I've carried you for nigh on to four years. Never knew till now that I'd killed anyone with you, though I've used you a lot. But this night you've killed a Yankee Colonel."

Tugging at Root's fine boots, he speaks to the rifle again.

"I reckon that's enough glory fer either of us."

Custer hasn't broken Walker's line but, west of the village, his cavalry now are astride the Lynchburg road. If they can hold their line in the morning, Lee's retreat will be blocked.

SHERIDAN

Sheridan makes his headquarters near the railroad station. Sitting on a rough wooden bench, he reports to Grant:

> Cavalry Headquarters
> April 8, 9:20 p.m.
>
> General,
>
> I marched early this morning...on Appomattox Depot....A short time before dusk, General Custer, who had the advance, made a dash at the station, capturing four (*sic*) trains of supplies with locomotives....
>
> Custer then pushed on toward Appomattox Court House, driving the enemy, who kept up a heavy fire of artillery, charging them repeatedly...
>
> If General Gibbon and the V Corps can get up tonight we will perhaps finish the job in the morning. I do not think Lee intends to surrender until compelled to do so.
>
> P. H. Sheridan
> Major General

He sends another courier for General Griffin. Griffin's infantry, that's what he needs now; by morning. The race is up to them.

Still a hard march away, General Joshua Chamberlain is awakened by Sergeant McDermott's hand on his shoulder, "Courier comin', sir. Orders I think."

Chamberlain struggles to wake, feeling the pain returning to his battered body. Then the sergeant thrusts a hot tin cup of would-be coffee into his hand. That helps.

The courier, from Griffin, has a copy of Sheridan's message to Griffin:

> I have cut across the enemy at Appomattox Station, and captured three of his trains. If you can possibly push your infantry up here tonight, we will have great results in the morning.

"Shall I have the 'General' sounded, sir?" McDermott interrupts Chamberlain's reading. Without reading Griffin's added orders, penned to the bottom of Sheridan's message, Chamberlain nods.

"Yes. Get the men up. We'll march in one hour. Appomattox Station. And, Sergeant Mac, have the bugler sound 'Officers Call'."

Less than an hour later, McDermott stops Chamberlain with one foot in Charlemagne's stirrup to persuade him to dismount for a tin plate of nondescript food and another cup of muddy coffee.

As Chamberlain rides to the head of his division, he thoughtfully gauges the men Charlemagne passes: tired, stiff, footsore, hungry, and cold. But their faces are flushed with excitement; no need to be told, they see the end in sight.

_____

Two hours later, closed up and marching hard behind Ayres' division, they're near Appomattox Station when Griffin's long column is intercepted by an excited courier who came straight from Sheridan.

"General, do you command this column?"

"Part of it; V Corps. General Chamberlain."

"Sir, General Sheridan wants you to break off and come to his support. Rebel infantry are pressing him hard, and our men are falling back. He says you're not to wait for orders through regular channels but to come at once."

Once more fate is thrusting the former Professor of Rhetoric and Languages into a critical fight.

THE PURSUED

Through the night and the following day Lee's army continued west, above the Appomattox, following deeply rutted country roads and crossing rough fields and soggy marshes that cause shoddy, green-leather soles to separate from their uppers.

Their passing is marked by more abandoned cannon, by wrecked and overturned wagons, and by men and horses left behind because they simply can't go any farther. Men with tattered, muddy clothes; eyes that are sunken and listless; faces that are peaked and pinched. Soldiers who've fought valiantly for four years now are dropping out from sheer exhaustion. More muskets have been left beside the road or their bayonets driven into the ground.

_____

Gordon and Longstreet, however, continue to have unarmed men pick up those muskets; and Lee's stronger veterans encourage the others to go on another mile or two.

Gordon hears those Gray veterans insisting that soon now Lee will turn them about, and they'll fight again, and somehow win again. And then it will be all right.

_____

Reverend E. H. Harding, Chaplain of the 45th North Carolina Infantry Regiment, is carrying a rifle now. He'd picked it up at Sayler's Creek, explaining that, "This is not the time for noncombatants."

_____

As the afternoon wears on, the march surprisingly seems a little easier for them, just as it does for Grant's soldiers. The road is firmer; the countryside is more beautiful; and the sun warms their faces. And their enemy seem to have vanished. Humphreys' men, who've pressed them since Amelia Springs, are back there but don't seem to want to bring on a fight.

Union cavalry General Gregg, taken in the fight below Cumberland Church, is among other Northern prisoners being held beside the road when a ragged column of Georgia infantrymen passes. He's wearing a splendid general's uniform, his hat adorned with a large, black ostrich feather.

From the passing ranks a Confederate soldier, wearing a tattered, gray, long, swallow-tailed frock coat and a butternut scotch tam, suddenly reaches out to snatch the fine, black ostrich-feathered hat from Gregg's head and, in the same motion, replace it with his own battered tam.

The smooth theft speaks of practice on other marches for the soldier doesn't break his stride or look back during the exchange.

Gregg angrily tears the scotch tam from his head and stamps it into the mud. There's not much more that he can do, however, except watch his own fine hat disappear up the road.

Foraging, if there were time for it, would be better in this part of Virginia largely untouched by the war, but there isn't time for soldiers to wander from their regiments.

Colonel Charles Venable and another officer of Lee's staff approach an old farmer and, embarrassed, ask if they might buy two of his hens for $25 Confederate money, as that's all they have. The old man sells the hens but won't accept that much money, Confederate or not, explaining, "Gentlemen, I've lived here sixty-five years, boy and man, and I've never asked more than a shilling (less than seventeen cents) for a chicken in my life."

Major Edward Boykin is watering his horse at a small stream when he sees an impressive Confederate officer on a large, gray horse approaching. He knows the rider, Captain Allen of the 24th South Carolina Cavalry Regiment, who bears such a striking resemblance to Robert E. Lee that he's often mistaken for the Confederate general.

Boykin starts to wave to his friend. Then his hand stops, drops to his side, and Boykin straightens in his saddle. This is not Allen but Lee himself. No escort, no staff; all alone. Erect, disciplined, well groomed, confident, and calm. It's the closest Boykin has been to Robert E. Lee, and his hand and arm stiffen as he salutes and reports, "Good morning, sir."

Lee smiles, returns the salute, and continues alone to the next rise. There he stops, turns in his saddle, and thoughtfully gazes south toward the Appomattox. The river is little more than a creek here, no longer an obstacle to Yankee cavalry or infantry.

Then Lee rides on, down the far side of the hill, beyond Boykin's view.

Near New Store, Lee confirms verbal orders he'd given the night before. He's reorganized his army, placing all its cavalry under Fitz Lee and its two corps of infantry under Longstreet and Gordon. Generals Pickett, Anderson, and Bushrod Johnson are relieved of their commands, ordered to return to their homes to await President Davis' call. Ostensibly because they no longer have men to command, with nothing said about Pickett's action at Five Forks or the three generals fleeing their commands at Sayler's Creek two days before.

Anderson will leave the army; Johnson and Pickett, unknown to Lee at that time, will trudge on with the army but without a command.

While he rests beneath a large pine beside the road, Lee is approached by Brigadier General William N. Pendleton: a fellow West Pointer from the Old Army, an Episcopal minister, and a close personal friend, now Lee's Chief of Artillery.

"Have you a moment, General?"

"Of course. How are you today?"

"I am well, sir; and you?"

Lee smiles gently, "Toler'ble, General, toler'ble. But better than yesterday. Those people seem to have fallen back, and though they're moving west below the river we have a good chance of getting to the railroad before them. Our rations will be there."

"The men are very tired, General. And hungry. You've seen them scouring the fields for anything they can find. The only meat that General St. John's people have left is 'Nassau bacon': salt pork, spotted like smallpox, awful smelling stuff. Have to chew it a long time or get it down fast."

Then, realizing that he's only revealing his own discouragement, Pendleton stops.

"I know," Lee's patience is immeasurable. "He issued the last of it this morning. Not a lot, but something. A few more miles," he smiles, "and we'll have food. I'm sure of it."

"General," Pendleton's still not sure how to begin, "I've been asked to speak with you. Yesterday several of our senior officers and I discussed our present situation. We mean no lack of faith in you, sir, but merely to share our views that you might not have to make some very painful decisions alone."

Lee says nothing but nods encouragement.

"We considered three possibilities, sir. First: to disband, scatter, reassemble later at a point you would designate. Second: to cut our way through the enemy, abandoning our trains that we might march faster, and continue our march. Third: to surrender. For us to disband and scatter would be very difficult, and we'd lose most of our men and equipment. The same with abandoning our trains."

"We decided, sir," he continues, "that we've done all that we can do and that our best recourse now is to surrender. As soon as possible, under the best terms we can get. I was asked to bring that recommendation to you, and so I have."

"And Generals Gordon and Longstreet?"

"They were not present at our meeting, and when I spoke with them about it this morning neither endorsed our recommendation. As a matter of fact General Longstreet, well you know 'Old Pete', reminded me that the Articles of War provide for officers or soldiers counseling surrender to be shot. I suspect that he means it. Both asked to be present when I spoke with you about this matter. Unfortunately, we are quite spread out, and I felt that it would take too long to get them here for this meeting."

It doesn't take Lee long to answer, and his voice is firm, uncompromising.

"Thank you, General. I appreciate your wanting to ease the burden I'm under. But we're not yet down to surrender. Not yet. We still have too many brave men to think of that. They still fight with great spirit; you saw that at Cumberland Church. Besides, if I even hint to General Grant that I'm considering surrendering, he'd take it as such a sign of weakness that he'd demand unconditional surrender. And I'd rather die than inflict unconditional surrender on these brave men, putting them at the mercy of our enemy. No, we must all be determined to die at our posts if Providence wills it."

Pendleton, embarrassed, seeing the hard iron beneath Lee's kind surface, reassures him, "Sir, we meant only to ease your burden. We're all perfectly willing to abide by your decision and, if dying is necessary, then to cheerfully die with you."

By late afternoon most of Lee's army is closing on Appomattox Court House, and Lindsay Walker reports that his reserve artillery train, far in front, already has passed through the village and now camps near the station where it's reported that four trains with provisions and clothing are waiting.

Lee establishes his headquarters beneath a large, white oak, about a mile east of the village and a hundred yards above the Lynchburg Stage Road.

The road climbs steadily from his woods camp to the village of Appomattox Court House. Then it continues west to Lynchburg, twenty-four miles away. On either side of the village open fields descend into valleys then rise again to long ridge lines, causing Appomattox Court House to sit much like a cup in a saucer, a little rise in the center of a broad valley.

Near the village there is a lot of confusion as Lee's arriving units, twisted and piled up, try to sort themselves out. Wagons and artillery pieces are crowded helter-skelter. Cavalry and infantry are restlessly moving about, uncertain, as if waiting for orders. The noise of Confederate bands, mustered near the road to cheer the tired men, only adds to the confusion.

An officer hails a group of about 250 Southern infantrymen, "What regiment is that?"

One, with a wry smile, calls back, "We're Kershaw's *division!*"

Then a Confederate brigade arrives: eight men clustered around its flag.

Colonel Henry Peyton, Lee's Inspector General, positioning a skirmish line south of the village, asks, "What command is this, Colonel?"

"We're all that's left of the 1st Virginia Infantry Regiment, sir. About two hundred of us, give or take, and we're supposed to protect the whole left flank of the Army of Northern Virginia!"

Doctor John Claiborne, passing through the village to reach the railroad station, sadly shakes his head at the decimated units he sees. Then he asks Peyton.

"Colonel, does General Lee really know how few soldiers are left and the shape they're in?"

"No, I don't believe that he does."

"Then whose business is it to tell him, if not his Inspector General's?"

"I can't tell him, Doctor; I can't."

The sun is setting when Lee, in the valley east of the village, hears the first heavy rumbling to the west. He immediately identifies it: cannon fire. Then the rumbling is followed by the harsh, distant crackle of rifle fire. A short burst, then silence, then volley fire, closer to the village. Custer's initial attack. Lee knows, without being told, that somewhere up ahead the ration trains, their last chance for food, are under attack.

Drummer boy Johnny Woods, west of the village, has found supper in a farmhouse. When he hears the firing behind him, he hurries to finish his meal. His host, however, urges him to stay.

"Lee's army's surrounded, boy. There's nothing that you or anyone else can do to help. He's gonna have to give up. Stay here with us; we'll hide you from the Yanks. When it quiets down, you can go home."

"No," the boy answers. "Thanks, but I can't stay. If it's all gone up back there, I guess I'll try to get through to Johnston's army. If I can't make it to North Carolina, I'll go home."

Along with other stragglers and disorganized Confederate cavalry, he begins to walk toward Lynchburg. That night he finds an abandoned horse, rides him into a cavalry camp, then shares their food and fire. A little later he falls asleep beside his drum.

THE GENERALS

In the evening, after the firing beyond the village has died, a courier finally reaches Lee with Grant's second message. Union General Seth Williams had difficulty passing it through the Confederates' skittish rear guard.

Captain Marcellus French, 35th Virginia Cavalry Battalion, watched as four Federal riders advanced from some woods. Then a Confederate marksman fired, and one of the Blue horsemen dropped while the others fled. French galloped ahead to find a small, no longer white flag beside the fallen man. A flag of truce, too small to be seen by those who shot its bearer. Then one of the surviving Federals hailed him.

"Flag of truce! Flag of truce!"

"Bring it here," French called back.

One of the riders, pale and nervous, rode forward, calling as he came, "Hold your fire! You Johnnies have shot three of our couriers today. General Williams back there has a dispatch for General Lee."

Then Williams arrived, accusing, "Captain, your men have fired on my flag of truce several times today. I've had men killed trying to deliver dispatches to General Lee."

"Captain French, General. Sorry; but that's the first flag of truce I've seen today, and I didn't fire on it. I'll pass the word, but that's a small flag. Didn't see it myself till I stood over it."

"I'm General Seth Williams from Grant's staff," the Federal officer warmed. "I have a letter for General Lee. Can you deliver it?"

"I can get it to General Rosser, sir."

"That will do," he handed the letter to French. Then he produced the same small flask of brandy he'd offered the night before to Confederate Colonel Herman Perry.

"Brandy, Captain. Have a drink."

French, not as skittish as Perry about such courtesies, took a long drink, thanked Williams, and turned to his own lines.

In his woodland camp Lee reads Grant's second message by candlelight:

April 8, 1865

General R. E. Lee
Commanding C. S. A.
Your note of last evening in reply to mine of (*sic*) same date, asking the condition on which I will accept the surrender of the Army of Northern Virginia is just received. In reply I would say that, peace being my great desire, there is but one condition I would insist upon, namely: that the men and officers surrendered shall be disqualified for taking up arms again against the Government of the United States until properly exchanged. I will meet you or will designate officers to meet any officers you may name for the same purpose, at any point agreeable to you, for the purpose of arranging definitely the terms upon which the surrender of the Army of Northern Virginia will be received.

U. S. Grant
Lieutenant General

As he reads Grant's message, Lee sighs in relief. Seeing the word "surrender" makes his choices all the more painful, but he's greatly relieved at the words "taking up arms again...until properly exchanged." Grant's committing himself to treating captured Confederate officers and men as prisoners of war, not as criminals; and he's not asking for unconditional surrender. Further, if Lee's troubled by the humiliation of having to personally surrender his army, Grant's suggesting that he give that job to a subordinate.

Well, I won't do that, Lee thinks to himself, but it's another indication that this is a sensitive man.

Without consulting anyone else he prepares his answer and hands it to an aide.

"Colonel Marshall, please get this to General Grant as quickly as possible."

Then he turns to the immediate situation his army faces. The light is failing quickly now, and he still does not know exactly what has happened beyond the village. Cannon and rifle fire, louder in the night air, still sound from beyond the ridge. It swells, subsides, swells again, much louder this time and more sustained. Finally, when gunners no longer can find their targets, it ends.

Around nine o'clock Pendleton brings him a firsthand report.

"General Lee, I was checking General Walker's artillery when enemy cavalry struck at the railroad station. Four trains were waiting; at least three of them have fallen to the enemy. I believe that the last train escaped. General Walker called on General Gary's cavalry to delay Sheridan's cavalry, Custer I believe, while he built a stronger line nearer the village. When I left, he'd repulsed the first enemy charges there. I've brought back perhaps sixty guns, but I fear that the remaining thirty or so have been taken by the enemy. I don't believe that Yankee cavalry can take the village, but they're across the Lynchburg road. At least two divisions of them."

"Have they infantry, General?"

"Not that I saw."

"Thank you, General," Lee's face is pale, weary, but he remains calm.

"Colonel Taylor," he directs, "have Fitz Lee move his cavalry to the front. Tell him to drive the enemy away if possible; if he cannot force them back, he must help General Walker hold his position. General Gordon should move his infantry forward also. Then ask Generals Gordon, Longstreet, and Lee to counsel with me here."

WAR COUNCIL

Near midnight the four generals meet at Lee's headquarters in the little patch of woods above the road. It's a simple camp with no tents, no tables, no chairs.

Longstreet sits on a log, smoking his pipe and, as is his way, listening or simply nodding his head in agreement or disagreement. John Gordon and Fitz Lee share a blanket thrown upon the cold ground. Lee stands before the flickering campfire, its flames illuminating his face. His hands, unless he is gesturing to make a point, are clasped behind him. Other staff officers and commanders are nearby, but only the four generals share this war council, the only one Lee has held since he took command of the army two long years before.

Watching a very tired, very old commander, Gordon, whose own mind is numbed at the latest crisis they're facing, knows the agony that Lee must feel. Heartbreak is written all over the Confederate commander's face. Still, Lee is calm, steady, reasoning; a commander Gordon knows the army still is ready to follow to the death if necessary.

Lee reviews the tactical situation: "It appears that at least a division, perhaps two, of General Sheridan's cavalry have seized the railroad station, and we've lost our ration trains. They then advanced on the Court House where General Walker and our cavalry and infantry have stopped them. They've fallen back now, but are astride the Lynchburg road."

"I don't believe that their infantry can have reached us yet," he continues, "but by morning they could be close."

Then he describes his own army. "We have perhaps 10,000 men with weapons who can be counted on to use them. Our supply of artillery pieces and ammunition is low. Our cavalry is very limited in size, and our horses and men are very tired. Our soldiers are very hungry but, if we can break through their cavalry, they are capable of another forced march to Lynchburg where rations surely must be waiting."

"I have received," he shares for the first time, "two messages from General Grant, requesting the surrender of this army. His terms, he promises, will be that our officers and men lay down their arms and promise not to take them up again until they are properly exchanged."

"My greatest fear," he reflects, almost to himself, "has been that, if our cause fails, our brave soldiers might be treated as criminals to be prosecuted rather than as honorable prisoners of war. General Grant's message implies that that will not happen."

"It seems," and he looks to Longstreet who knows Grant better than any of them, "that this is not the same General Grant who two years ago offered General Buckner (Major General Simon Boliver Buckner) at Fort Donelson only unconditional surrender, and that to be accepted very quickly or he would attack. But then," he smiles, "we all have changed in that time."

Longstreet, Grant's long-time friend, nods his head and mutters around his pipe, "Grant'll do the right thing, General. We can depend on that."

Lee nods, thankful for Longstreet's reassurance. He does not share with them Grant's offering him the option of having a subordinate arrange the terms of surrender. He appreciates Grant's thoughtfulness, but he's not otherwise considered the suggestion. Nor does he discuss with them his earlier conversation with General Pendleton, suggesting that he surrender. Very weary, weighed down by the burden of the decisions only he can make, he still is very much in command of the army. And, if it must be surrendered, he'll not pass that decision or that duty to anyone else.

Then, extending his hands, palms up toward them, he concludes, "Gentlemen, I would appreciate your views on the options before us."

When he's finished, they are all silent for a moment, the only sound the crackling of wood burning in the campfire before them and, off in the distance, an occasional rifle shot by a nervous picket, loud in the night air.

It's Gordon, the most articulate of them, who begins.

"Sir, it seems to me that we're down to three choices. First, disband our army. Tonight. Abandon our heavy ordnance and trains in hopes that we can find more later. Order our men to find their own way through the enemy lines, to meet again wherever you designate, perhaps Lynchburg; or," and he hesitates to suggest it, "to fight as partisans until such time as the army can be reconstituted."

"Our second choice," he hurries on, "is to mass our remaining strength west of the village and attack early in the morning. Cut our way through the cavalry before us, then take up the march to Lynchburg."

When he's seen Longstreet's reflective nod around his pipe, he turns to their third choice.

"Our third choice, I believe, is to surrender. Now, while we still can negotiate terms. Another day without food and allowing our enemy to arrive in full strength, and we'll not be in position to bargain anything."

Lee nods then, "Thank you, General Gordon."

Then he turns to the others, "Are there any other options open to us?"

When there are none, Lee takes up the discussion, personally ruling out the first option.

"If we disband now, we probably can never gather our army in any strength at all. Our men have been very loyal, but we'll lose a great many of them if we dismiss them this way. And if they abandon their equipment, even if we rally later, they'll not be able to continue the fight. Most important of all, however, if our soldiers turn to guerilla warfare, it can only bring years and years of ruin upon our country. It will cause our brave soldiers, who've fought honorably and well for four years, to be branded criminals. Apt to be hanged on the spot. Following the war Federal soldiers would garrison our towns, searching out fathers, sons, husbands believed to be fighting as partisans. All that can only bring more retribution upon the South. No, disbanding our regiments and ordering our officers to fight alone or in small groups, is not an acceptable solution."

They agree, and that narrows it down to fight or surrender. Attempt to break through whatever's out there then march toward Lynchburg with the vague hope of then swinging south to join Johnston's army; or negotiate the best terms they can before their bargaining position deteriorates even more.

The decision, their experience tells them, rests upon whether they still have enough strength to break through the enemy blocking the Lynchburg road.

Gordon and Fitz Lee, with nothing but instinct to back their belief, are convinced that Federal infantry are with Sheridan's cavalry west of the village or will be there by morning. "Already," Fitz Lee points out, "we can see their campfires on three sides of us."

Robert E. Lee doesn't agree, shaking his head, "I do not believe that their infantry have marched that quickly but, if they are not there now, it's true that they will be there soon."

Longstreet, the pragmatist, takes a last puff of his pipe then takes it from his mouth and uses it to emphasize his thoughts, "Gentlemen, we don't know that Grant's infantry's out there. If it's only cavalry, we can break through. If infantry's there, well, we'll have to face that when it comes. It seems to me, that we have no choice 'cept to try."

So they agree. At dawn they'll try to cut their way through whatever's blocking the Lynchburg road ahead of them.

"General Lee," the decision having been made, Lee turns to his nephew Fitz Lee, his voice stronger, firmer, "your cavalry, supported by

General Gordon's infantry, will lead the attack. If you can break through their lines, continue to push them back, holding the road open until our trains and General Longstreet can pass through. General Longstreet will continue to protect our rear while General Gordon again takes up the advance."

"Time, sir?" Gordon asks.

"Four o'clock. That will not allow our men much rest, but we have no choice."

Before their council ends, Fitz Lee mentions the unmentionable, "Sir, if our attack fails and you are forced to surrender the army, please try to warn me beforehand that I still might escape with the cavalry."

Lee stares at him for a moment before answering. The act of ordering his army to attack, as he'd done successfully so many times before, has buoyed his spirits, but Fitz Lee's question causes him to face the desperate nature of this attack.

"Yes, if their infantry are present and our attack cannot succeed, then I must face the inevitable. If that occurs, and I have time to do so, I will release your cavalry."

Lee, the gambler, will gamble one more time, and the survival of his army, probably of the Confederacy, will depend upon their success.

———————

The war council over, the generals turn to return to their commands. John Gordon has barely left Lee's campfire, however, when he realizes that he'd forgotten to ask Lee where he should make camp the next evening, a task routinely assigned to the lead unit on a march. He sends an aide back to inquire.

Lee, deep in thought beside the fire, hears the question and gently smiles.

"Please tell General Gordon that I should be glad for him to halt just beyond the Tennessee line."

The Tennessee line, they all know, is nearly two hundred miles away.

GRANT

Grant, despite frequent applications of hot and cold moist cloths across his forehead and soaking his feet in hot water laced with mustard, is suffering from the migraine headache that plagued him during the day then worsened during the evening. Finally giving in to it, he lies down on a horse hair sofa in the sitting room of the little farmhouse.

A dozen of his staff are scattered throughout the house, wherever they can find room to spread their bedrolls. Several younger officers, too excited to sleep, find a battered old piano and begin to bang on it then to sing. Louder until Grant, in agony, sends word that they must stop.

At midnight the quiet is interrupted by the challenge of a guard outside. Then the door is thrown open and a cold, tired Colonel Charles Whittier from Humphreys' staff enters.

"Dispatches for General Grant."

Chief of Staff John Rawlins tears open the envelope, hurriedly reads its contents, then goes to Grant's room.

Hoping that Grant has been able to sleep, Rawlins opens the door quietly, tentatively. No need. Grant has not slept at all.

"It's all right, Rawlins. I'm awake. No one could sleep suffering this way. Is it from Lee?"

"Yes, General," the others hear the anger in Rawlins' voice.

Then there is silence as Grant reads Lee's reply.

8th April '65

Genl

I recd at a late hour your note of today. In mine of yesterday I did not intend to propose the surrender of the Army of N. Va. - but to ask the terms of your proposition. To be frank, I do not think the emergency has arisen to call for the surrender of this army, but as the restoration of peace should be the sole object of all, I desired to know whether your proposals would lead to that and I cannot therefore meet you with a view to surrender the Army of N. Va. - but as far as your proposal may affect the C. S. forces under my command & tend to the restoration of peace, I shall be pleased to meet you at 10 a. m. tomorrow on the old stage road to Richmond between the picket lines of the two armies.

Very respy your Obt. Servt.
R. E. Lee
Genl

Then the listening staff officers hear Grant's voice, more disappointed and saddened than angry, venturing, "Well, it looks as if he means to fight."

Rawlins, however, is too worked up by Lee's message to pass it off that lightly.

"He did not propose to surrender!" Rawlins mocks Lee's words. "He *did* propose to surrender. He knows it; you know it; we all know it. And he says 'tend to the restoration of peace.' Be careful, Grant. He's trying to trap us into a general treaty of peace. You didn't offer that. Said nothing about a general treaty."

"You asked him to surrender," Rawlins storms on. "When he asked your terms, you gave them: lay down your arms until you're properly exchanged. That was very proper. But now he's after something far beyond the surrender of his army. Something that the President reminded you in February is beyond your authority."

"His reply," Rawlins continues, "is a positive insult, Grant; an attempt to change your whole purpose in writing him."

"Now, Rawlins," they hear Grant's quiet voice. "They're only words; he's trying to be let down easy. We'd do as much. If I can meet him in the morning, we can settle this whole business within an hour."

Rawlins, however, won't be appeased. Again he mocks Lee's message, insisting that Lee's only trying to buy time and better terms. Grant, however, again quietly defends the Confederate general.

The staff, listening beyond the little room, can almost see Rawlins, pacing back and forth, shouting his recriminations while Grant (his head, they agree, must be splitting) calmly waves his hands and suggests, "It's all a tempest in a teapot, Rawlins."

Then Rawlins storms again.

"He doesn't think that 'the emergency has risen.' Why the emergency's been staring him in the face for forty-eight hours! Well, Grant, if he doesn't see it, we must help his comprehension. He *shall* surrender. By the eternal, it shall be surrender and nothing else!"

Then Grant's quiet, "Now, Rawlins; Lee's in a bad fix; but he's still a soldier and has to follow his government's wishes." Then he returns to his basic belief, "If I can meet with him, he'll surrender."

Rawlins, realizing that he's not going to change Grant's thinking, fires a final warning.

"If you meet with him, Grant, remember that you have no authority to discuss a general peace. Your job's to capture or destroy Lee's army. You can meet with Lee but only to accept the surrender of his army. Nothing more."

A long silence then they can almost hear Grant sigh, "All right, Rawlins. You've carried your point. As usual. I'll answer his letter in the morning. Now let me try to rest."

Later, when Colonel Horace Porter quietly goes to see whether Grant's been able to sleep, he finds the sitting room empty. And Grant's nowhere in the house. Porter, anxious, finds him pacing the farm yard, holding his throbbing head.

They try more home remedies but nothing helps. Then Porter persuades him to go to Meade's headquarters for a little coffee. When he's done that and feels a little better, Grant sits at a small table to draft his third message to Robert E. Lee. While he's doing that, they alert General Seth Williams. He's about to make another dangerous ride.

LINCOLN

At City Point this is Abraham Lincoln's last day with Grant's soldiers. He spends most of it in the telegraph office, playing with the three fat little kittens and reading the latest dispatches from Grant and Sheridan. All continues to go well.

Then he goes to the Depot Field Hospital to visit wounded soldiers. Hospital officials begin to brief him on its operations, but he smiles and waves them aside.

"Gentlemen, I'm perfectly content that you know better than I how to conduct these hospitals. You needn't tell me of all that. I'd just like to take the hands of the men who've given us these glorious victories. Show me our soldiers, gentlemen. Let me visit with them."

At the VI Corps' hospital patients who can stand instinctively form ranks before their wards to greet their President. Lincoln will have no military formality, however, simply stopping to shake each man's hand and to share some personal greeting.

"Are you well, sir?" "Are they taking good care of you, son?" "And where are you from? Really? Why, I've been near there myself. I recall...." and he shares an experience sure to make them laugh.

Then he enters the wards, wanting to meet those who'd not been able to come outside.

He pauses before one cot, extending his hand.

The patient takes it, grasps it firmly, then asks, "Mr. President, do you know who I am?"

"No, sir," Lincoln's sad smile warms the room, "I do not."

"Well, sir, you've offered your hand to a Confederate colonel who fought you as hard as he could for four years."

"Well," Lincoln nods, "I'm glad that Confederate colonel didn't refuse to take my hand nor to offer his."

"No, sir," the wounded Rebel grins and, when the President has gone, confides to the others, "He had me whipped from the moment I saw him."

Back at the *River Queen*, the crew is preparing the ship for its Washington run late in the evening. Admiral Porter, hearing more rumors of an assassination attempt, has rechecked every member of the ship's crew and personally briefed its officers on their responsibilities.

Lincoln has shaken the hands of perhaps 5,000 wounded soldiers this day and, returning to the *River Queen*, his hands are badly swollen and he's very tired. Mary Lincoln, however, has arranged a final dinner aboard the steamer, and he must attend. Porter, accompanying the President, notes that Lincoln seems weary but content.

After their dinner a military band comes aboard for a farewell concert. Asked to select a piece, Lincoln has them play the "Marseillaise," in honor of the Count de Chambrun. Then, much to the surprise of everyone, he calls for a tune he tells them he particularly likes: "Dixie." As the musicians fill the ship with the stirring strains of the Southern anthem, Lincoln confides to Porter, "That tune is now *Federal* property, Admiral, wouldn't you say?"

The party ends, and just before midnight the *River Queen* noses out into the dark James River. Most of the obstacles are gone now, but the ship proceeds slowly, carefully toward the bay and the ocean beyond.

When the ship casts off, most of the revelers are asleep. Abraham Lincoln, however, stands alone near the rail for a long time, absorbed in his thoughts, gazing at the dark banks slipping by.

Then he turns toward his own cabin; tomorrow he must pick up the load he'd largely been able to set aside for a little while. Perhaps, when they reach Washington, there will be a message from Grant. The message he's prayed for for four long years.

DAVIS

In Danville, a hundred miles below the little courthouse village where Lee's soldiers are preparing for their desperate attempt to break through the Northern lines, Jefferson Davis has learned a little more about Lee's circumstances.

He's received a report from Vice President Breckinridge, telegraphed from a small station up the Richmond-Danville line. It relates Breckinridge's conversation with Lee at Farmville. Then Breckinridge describes Lee's terrible losses at Sayler's Creek and his own impression of a defeated, staggering army trying to outrace their Federal pursuers.

"He'll try to move toward North Carolina," Breckinridge concludes, "but," he adds, "the situation is not favorable."

After dinner Davis is talking with several cabinet members when an aide enters to tell him that a very tired courier has arrived from Lee's headquarters.

Ushered into the dining room, the courier is bone-weary Lieutenant John Wise.

The young officer normally would have felt self-conscious in such grand company. His uniform is torn and dirty; his face is dirty, scratched, beard-stubbled. But he's been through too much for that to really matter. Davis and the others listen attentively to his report.

Finally a cabinet officer asks the most important question.

"Lieutenant Wise, do you think that General Lee will be able to escape with his army?"

The young officer's mind goes back over the hazardous trip he's made, the things he's seen: his father draped in an old saddle blanket; men counting on food that never arrives; Lee: old, tired, still unconquered, refusing to accept food himself until his men are fed. Thinks of all of it then shakes his head.

"I'm sorry, sir, but no, he cannot. From what I saw and heard, I believe that General Lee must surrender."

His listeners instinctively shudder. Then he continues, "Mr. President, General Lee's surrender is only a question of a few days at the most. The result is inevitable, and postponing it only means the useless loss of more noble, gallant blood. I think that he'll know that and do what he must do."

Then he concludes, "I'm sorry, sir, but you asked. If it has not yet happened, General Lee will be forced to surrender his army. He has no choice."

CHAPTER 7

Appomattox Court House   Palm Sunday, April 9, 1865

LEE

As Grant, unable to sleep, paces a farm yard a dozen miles from Appomattox Court House, Colonel Charles Marshall, Lee's secretary, so nearsighted that he sleeps with his spectacles across his nose, has pulled his overcoat over his head and is trying to make the best of a cold bed on the hard ground. Cold, sore, and fearful of a Federal attack on their camp, he can't sleep.

He glances about their headquarters, a little camp in a pine and oak woods' clearing above the Richmond-Lynchburg road. Outlined in the flickering light of a faint campfire, it seems to Marshall that the camp is more like that of a rifle company commander than that of the commanding general of all the Confederate armies. Officers and men, most without tents or wagons, like him, are rolled up in their saddle blankets or in their greatcoats and sleeping on the open ground.

Suddenly he becomes aware of a new sound below the sleeping camp. An alarming sound: that of marching men. It's muffled by the overcoat around his head but clearly is close by. He draws a large pocket watch from his pocket, and uses the campfire's light to check his watch: 1 a.m. Then he fixes the sound: it's on the stage road below; and he recognizes it: the steady tread of boots and the creak and rattle of rifles, bayonets, and canteens. Marching men.

Then those sounds are lost in the early morning low, rhythmic chant of an easily recognizable Texas' marching song:

The race is not to them that's got
The longest legs to run,
Nor the battle to that people
That shoots the biggest gun.

Marshall can't see the men, but he knows from their chant that what's left of Lieutenant General John Bell Hood's Texas Brigade are passing, singing as they lean into the steep road that leads to Appomattox Court House.

Hood's Texans, legendary for their bravery on so many fields, their ranks shattered on the rocky slopes of a little hill just south of a Pennsylvania farm town nearly two years before. Only a few hundred of them left, but this cold night called forward again to help John Gordon break the Federal lines above them.

Relieved, Marshall again looks about their camp. It's a poor camp, he admits. No headquarters' tents, only a wagon or two to offer some shelter from the weather. Lee's own wagon bears the markings of a New Jersey regiment. The few tables, chairs, and blankets they have left also bear Federal markings.

"This army," Marshall sighs, "truly is down to the last ditch." Somehow comforted, however, by the soldiers' chant, he drops back for a few more minutes' rest.

A bit later General Pendleton arrives, surprised to find that Lee has not rested at all and now stands warming his hands before the campfire. He's even more amazed at Lee's appearance: Lee is wearing a clean white shirt and an immaculate new Confederate general's uniform. His new boots are freshly polished, and he wears a broad, red sash about his waist. Buckled to his sword belt is an elaborately engraved sword. Lee, who invariably campaigns in a linen duster and slouch hat, this morning is magnificent in appearance.

Pendleton notes, however, that Lee's face, reflected in the fire light, is that of a haggard, very tired, very old commander.

"Good morning, General," Lee smiles and welcomes him with a wave of his hands.

"Good morning, sir. You certainly look splendid."

"Well," Lee smiles a bit ruefully, "it's the only clean uniform I have. And," he adds, "perhaps it's just as well. I'll probably be General Grant's prisoner today, and I thought it best to make my best appearance."

"Now," he continues, "you've spoken with General Walker?"

"Yes, sir; he's withdrawn with his remaining guns to the west, escaping Sheridan's cavalry. I've told him that you mean to attack in the morning. I doubt that he can support us with his fire, however, and I expect that the most that we can hope for is that he'll be able to continue west toward Lynchburg."

Lee nods. "We'll attack west from the village. If Fitz Lee can clear the road, Gordon will hold the shoulders open for us to pass. Then we'll follow General Walker."

"General Gordon," Lee sighs, "has about 3,000 men with weapons. A good sized brigade when this war began. Fitz Lee has another 2,400. General Longstreet has another 6,000 men behind his breastworks in our rear. In all, perhaps 13,000 men and your sixty guns. Gordon will face at least twice that number of cavalry on his front alone. Not very promising. And, if infantry are present, the game is over. You understand that?"

When Pendleton nods, Lee gently smiles and pats his shoulder.

"Get some rest, General. Wherever you can. But stay nearby should I need you."

General Lee mounts Traveller and rides forward, hoping to be able to gauge Gordon's attack for himself. The weather's not promising for that, however, as a heavy fog mantles the land. There's no sound ahead of him and Gordon, he knows, must begin his attack soon. Perhaps the fog will be a blessing, allowing his infantry with their single-shot Enfields to close on their enemy before they again must face Sheridan's repeating rifles.

THE LAST FIGHT

West of the village Sheridan also has done all he can to prepare for the fight he feels is inevitable. His cavalrymen have used fence rails, felled trees, and dirt to build breastworks across the road. Several hundred yards ahead of those breastworks they've posted cavalry pickets, dismounted, to delay the enemy attack as long as possible. Long enough, Sheridan hopes, for Ord's and Griffin's infantry to come up.

About 4 a.m. he meets with Generals Ord, Gibbon, and Griffin who'd come ahead of their troops to Sheridan's headquarters near the railroad station. They report that their divisions are strung out, but that their leading regiments are fast-closing behind them.

Sheridan, without authority to command the senior Ord, suggests that Ord's infantry block the road behind his cavalry's breastworks. Ord nods, assigns the job to Gibbon. Griffin will close on the station then move in on Gibbon's right.

At the edge of the village Confederate Generals John Gordon and Fitz Lee also are conferring. Nearly dawn now as they study the Union breastworks before them. Nearby, waiting for their orders, is Major General Bryan Grimes: a square-faced, broad-shouldered North Carolinian, a fiery commander.

"That's dismounted cavalry out there, Fitz," Gordon lowers his field glasses. "You take them, and I'll move in behind you."

"No, John," Fitz Lee makes his own assessment. "There may be cavalry there, but there's infantry too. Lots of them. I won't be able to budge them. Let me cover your flanks while you make the assault."

The two generals argue, their "after you; no, after you" discussion at last getting the better of the impatient Grimes.

"Gentlemen," he interrupts, "it's getting light. And the lighter it is the more trouble we'll have. Somebody has to do it, and do it now. Cavalry or infantry, we *can* drive them from the crossroads. I'm sure that I can do it."

Gordon, as aggressive a commander as any in either army, still delays. Seeing that, Grimes again angrily demands that someone do something or let him do it.

"All right," Gordon lowers his field glasses again, "you drive them off. Fitz, cover our right flank. General Grimes, when you've secured the crossroads, I'll swing across to hold the road open and take the lead. You follow behind us."

"Well, I can't do it with my men alone," Grimes answers. "I'll need help."

"Take as much of the corps as you need. Yours until you clear the road. Send word as soon as it's clear."

It's nearly 8 o'clock when Federal cavalry make the first move, charging the Confederate picket line facing Sheridan's breastworks.

Sheridan, watching the scarlet and white guidoned lines move forward, grins and calls to several Confederate surgeons, prisoners, "Now, boys, you're going to see something grand!"

One of the doctors quietly answers, "Perhaps. But I expect that they'll be back right soon."

He's right. Grimes' infantry easily breaks the Federal charge. Then Confederate skirmishers begin to move forward, Grimes' infantry hitting the center of Sheridan's picket line while Fitz Lee's cavalry drive in from the right.

As Gordon and Fitz Lee's men close on the Federal breastworks they're pounded by cannon and rifle fire. Some of them fall; the remainder, however, raise their Rebel yell and charge, breaking Sheridan's hasty breastworks and capturing several hundred prisoners and two Union cannon as Sheridan's men fall back.

"The way's clear, General," Grimes reins his horse to report to Gordon. "And the Yankee dead are all wearing cavalry uniforms. No infantry. If you can hold the shoulders, our army can pass."

Gordon, exuberant with Grimes' success, turns to order his infantry to advance on either side of the road.

Making one last sweep of the enemy before him, Gordon stops dead in his tracks. Standing erect in his horse's stirrups for a better view, he uses his field glasses to scan the road to the west. Then he swings the glasses to the south.

As Grimes has reported, Sheridan's cavalrymen have given way, opening the escape road. But, as the Blue cavalry part, Gordon sees solid lines of Union infantry behind them.

Federal infantry. No doubt about it. Waiting. Silent, ominous, waiting. So many battle flags. He identifies the units, counts them, is staggered by their meaning. Ord's Army of the James with Gibbon's corps and Birney's Negro division, and Griffin's V Corps. A triple line of infantry with red, white, and blue National Colors and several score of regimental, brigade, and division flags blossoming at neat intervals like bright poppies in a vast field of blue. As if calling for his attention, the flags silently flap in the morning breeze and the early sun reflects off thousands of bayonets and rifle stocks.

Federal infantry. Silent. No commands; no music; no shouts. Simply waiting. Lee's escape route is blocked. Bad enough, but Gordon's also seen a new threat on the Confederate left flank. More Union infantry are there; by their flags, part of Griffin's corps. And beyond them still more Federal cavalry are circling east toward the village. If Gordon advances at all, he's in great danger of being cut off and annihilated.

Stunned, he watches the massive forces gathering before them. Worse than he'd feared; far worse than he'd feared.

Then Colonel Venable of Lee's staff is there beside him.

"General," Venable's voice is soft, sympathetic, as if he knows the answer before he asks his question, "what shall I tell General Lee?" Gordon sighs then answers, "Colonel, please tell General Lee that I've fought my corps to a frazzle, and that I can do nothing more unless Longstreet can support me."

Venable nods, salutes, and wheels his horse about. Gordon watches him for a moment then turns again to the enemy before him.

"Colonel, catch up with General Grimes. Tell him to keep his lines intact, but to pull back toward the village."

As Gordon watches, however, Grimes' regiments do not pull back. Instead, they remain where they are, facing the overwhelming Federal force before them. Two more times Gordon sends couriers to Grimes. Three times Grimes refuses to obey. Finally, however, the combative Grimes begins to fall back toward Appomattox Court House.

Gordon, who'd watched Grimes' infantrymen's magnificent fight to clear the road, now watches their sullen withdrawal.

A Southern sergeant orders his men to retire but adds, "Don't let 'em see you running, boys."

A young Confederate officer has reached a corncrib when he decides that he'll not retreat another step. It's a personal decision. He doesn't bother telling anyone about it, and no one bothers to question it. Instead, Gray soldiers move around him, leaving him standing there, his pistol in one hand a sword in the other, waiting for the first Yank to appear. Determined to make a stand. A pebble on a very vast beach.

"Where shall I form my line, General Gordon?" an angry Grimes comes back with the last of his retreating men.

A weary Gordon waves his hand toward the village.

"Anywhere you choose, General."

"What does that mean?"

"It means that it really doesn't matter. We can do no more; we're going to be surrendered."

Grimes erupts again, "Surrendered! Surrendered! Why didn't you tell me? I could have gotten away while the road was clear. I still can, and take my men with me!"

He's turning away when Gordon reaches out to take his arm.

"General Grimes, do you think that you can feel one bit worse than I? Or General Lee? We've lost, sir. Face it! It's over. Over. And if you desert the army now, you'll stain your honor as a soldier. That will disgrace you, and it will disgrace General Lee. Consider your actions, sir."

Grimes, persuaded by Gordon's argument but still very angry, turns away. As he returns to his command, he decides not to tell his men of the inevitable surrender; not yet. He finds, however, that they've already guessed what is happening.

As soon as he's returned, one asks, "General, are we surrendered?"

"I'm afraid that we will be."

The man turns away and, tears streaming down his cheeks, throws his musket on the ground. Then he yells his defiance to anyone who cares to hear, "Blow, Gabriel, blow. My God, let him blow! I'm ready to die!"

Meanwhile, Colonel Charles Venable is reporting to Lee that Gordon's attack has failed and that Gordon is pulling back toward the village, unable to do anything more.

Lee nods calmly at Venable's report then quietly shares, "Then, Colonel Venable, there is nothing left for me to do but to go and see General Grant. And I would rather die a thousand deaths."

The fighting, however, still has not ended. Not if Sheridan has his way. The Confederates before him are pulling back, but Sheridan's not willing to admit that it's over. He has Ord's army and most of Griffin's corps west of the village with the remainder of Griffin's divisions hurrying in from the south. Even now Chamberlain, riding ahead of his troops, is reaching the little hill where Sheridan, astride Rienzi, is waiting. And beyond Chamberlain's brigade, Custer's cavalry are hurrying to get into position to strike at Lee's left flank.

Custer, Sheridan smiles to himself, his blood's up to smash them once and for all. And Chamberlain: thinks too much, but he's as good an infantry fighter as I've seen. No, he decides, I'm not done with them yet.

His face flushed with excitement and his whole being glowing with the lust of battle, dark and menacing on coal-black Rienzi, Sheridan cuts off Chamberlain's salute.

"Go in, General," he waves him forward. "Go in there and smash them."

Chamberlain nods, scans the field ahead of him. He's above a shallow valley that runs roughly west to east just south of the village. The field ahead slopes down to bottomland that crosses a small stream, Plain Creek. Beyond the stream the ground rises again to another ridge, perhaps six hundred yards away. The village is on that ridge. Before him are scattered clusters of Confederate cavalry, infantry, and some artillery. By their flags, Confederate regiments, but each so small that there seem to be more Southern battle flags than men to bear them. To his left, he sees the long, solid lines of Ord's and Griffin's infantrymen across the Lynchburg road. To his right rear Custer's men are closing on the village itself. A successful attack here would drive to the heart of the Confederate army.

One of Chamberlain's officers, making his own estimate of the situation, whistles, "Whew! Will you look at that! The devil's to pay, for sure."

A nearby soldier gives the riflemen's perspective, mocking their foolishness in marching so hard to get there.

"'Get to Appomattox Station, men! That's where your breakfast's waitin'.' So we've marched our legs off and gotten here. Just in time to be the last to die in this war."

Across the valley, Confederate infantry are crouched behind a stone wall halfway up the slope. Cannon are there too.

"No one," Chamberlain speaks his thoughts, "wants to make this attack; and no one over there wants it to happen."

Then he waves his arm and, down the slope and across the stream, his skirmishers advance.

They quickly come under the first, scattered fire, and men begin to fall.

Chamberlain turns to an aide, "Captain, have the batteries throw a few shells just this side of that wall. Maybe we can herd them back."

A Union artillery battery opens fire, and one of its first shells demolishes a ramshackle building just beyond the stone wall, near the center of the Confederate line. Dozens of frightened, clucking hens fly in every direction. Hungry Confederate soldiers drop their rifles to chase them, and the hens that cross the little stream find themselves being chased by Chamberlain's hungry Yankees.

"Sergeant Mac," Chamberlain grins, "tell the commanders to get our boys back into ranks. If we could find a big enough chicken coop, we could end this war without any more fighting."

McDermott grins, salutes, and hurries off.

The Confederate soldiers abandon the wall, slowly, sullenly pull back. Chamberlain's men cautiously follow, up the far slope to crest the ridge near the village.

Beyond them is another valley, perhaps a mile across, then, in the distance, a series of broken hills. A vast, relatively open amphitheater dotted with scattered fields and woodlots.

In the valley Chamberlain sees dusky masses of Confederate infantry slowly moving about, restless. In its center there is a small stream, a rivulet his men can jump across: the very beginning of the Appomattox. The chase, begun almost ninety miles away, now is down to this little stream. His bugles again urge his lines forward.

A Confederate cannon fires and its shell seems to make a direct hit on Private William Montgomery, Company I, the 155th Pennsylvania Zouave Regiment. Bits of uniform and equipment fly in every direction. Montgomery, it seems, has been blown into a hundred pieces. When the smoke clears, however, the "body parts" are only bits of his equipment and clothing. Montgomery, a teen-ager, will never be luckier.

A bullet pierces the shoulder strap of another soldier's cartridge box. The Minie ball, probably fired at long range, is stopped by the leather but leaves a bruise the size of a walnut on the soldier's chest. Another lucky Union soldier.

Not so lucky is Lieutenant Hiram Clark, 158th New York Infantry Regiment, struck by a shell fragment and one of the last Union soldiers to die.

Confederate soldiers continue to back away from their enemy. On his right, Chamberlain's men near the village. The sporadic firing continues there, with some civilians caught between the two armies.

Near the edge of the village, Private Henry T. Bahnson, the Confederate sharpshooter who'd personally captured several dozen Yankee soldiers in a railroad cut above Farmville, kills three of Custer's cavalrymen. Then, captured himself, he must wait beside one of the three he's slain.

Bahnson, a young, tired, frightened soldier, sees the slain youth's torn, bloody coat and white face, and begins to cry.

An older Yankee soldier, Sergeant Benjamin F. Weary, 2nd Ohio Cavalry Regiment, his canteen half-emptied of its whiskey, sees Confederate soldiers grouped around their flag near the courthouse and recklessly charges them alone, shouting, "Surrender, you damned Rebels! Surrender!"

He reaches the Confederate flag, seizes it, and is turning to gallop away when more than thirty Gray riflemen open fire. Struck by dozens of bullets, horse and man are dead before they reach the dirt street.

Doctor John Claiborne is taken prisoner by a Yankee cavalry sergeant who threatens to shoot him until Claiborne bribes him with a beautiful pair of spurs.

Custer's hand-picked honor guard, bearing more than forty captured Confederate battle flags, closely follow the Union general as he hurries up to strike the final blow.

When his regiments have wheeled their mounts into line, Custer takes his place before them, draws his saber. Behind him the "snick" of thousands of steel blades also clearing scabbards is sharp in the morning air. Buglers wet their lips, raise their bugles, prepare to sound "Charge!"

Then something happens across the Union front.

Several Confederate riders, each bearing a white flag, appear in the Confederate lines. The nearest, waving what seems to be a dirty white towel, passes through an opening in the enemy lines and rides toward Chamberlain.

Seeing the flags, Custer orders, "Hold up. Hold up. If it's a flag of truce, we can't charge."

Several more white flags appear, their bearers riding toward the Union lines.

"Hold up! Cease firing! Hold up!" the cry echoes across the field.

Both armies, thousands of men, Blue and Gray, watch the riders with their white flags. Men on either side of the last breastworks either army will face, stand dead still, hardly breathing. The only actors in the immense drama are the horsemen riding forward, waving their flags of truce.

Joshua Chamberlain, instinctively a professor once more, draws his watch. It's a little after 10 a.m., Palm Sunday, April 9. Then he snaps its cover shut and waits for the lone Confederate rider to reach him.

LEE

Lee knows from John Gordon's report that he can't break through the enemy before him. And now he's learned that they're also threatening his left flank. Even as he considers that he hears the boom of more cannon to the east. Meade's closing on his rear; Longstreet's already engaged.

For a moment, just a fleeting moment, he drops the mask of command, muttering to himself, "How easily I could be rid of this and be at rest! I have only to ride along the line and all will be over!"

Then he recovers, the temptation set aside for good, and sends for Longstreet. Longstreet brings along Generals Porter Alexander and William Mahone. Mahone, terribly thin and wrapped head to foot in a threadbare overcoat, goes to the fire to warm himself as the other generals begin to talk.

Lee tells them of the situation to their front and on their left flank; Longstreet reports being pressed by Humphreys' II Union Corps but that so far he's been able to hold their enemy back. Then Lee speaks of surrender.

"When I last asked you, General Longstreet, you said, 'No; not yet.' What do you say now?"

Longstreet takes a moment, lighting his pipe from a burning ember, then answers with a question, "Can sacrificing this army help our cause in any way?"

"I think not."

"Then, sir, your situation speaks for itself."

Lee, not surprised but still hoping for some other solution, turns to Mahone. Mahone, coming when called, is shivering badly and wants to be sure that the shivering isn't misunderstood.

"I don't want any of you to think that I'm afraid. I'm not; only chilled."

Lee smiles and takes Mahone's cold hands in his own. "That you might be afraid, General, would be our last concern."

He uses his worn linen map to brief Mahone on their situation. Then he asks Mahone's advice. Mahone agrees with Longstreet; Lee has no choice but to surrender.

Lee nods but again turns to Longstreet, "General, are you certain? Have we any alternative?"

Longstreet removes his pipe, shakes his shaggy head, and again answers, "No, General. We do not."

Lee then turns to Porter Alexander.

"General Alexander, do you agree?" Porter Alexander does not directly answer. Instead he tries to reassure Lee. "If you believe that we still can cut our way out, sir, I'll answer for my artillery. As we rode here my boys called to tell me that, if I have to abandon our trains, I'm to hang on to our cannon and shells."

"I know," Lee smiles, "they are good men. But we have only two small divisions left. Less than 15,000 men who can fight. There are four or five times that number before us and at least that many behind us."

"Then, General," Alexander answers, "I see only two choices: surrender or scatter our army. And the men, sir, would rather scatter than surrender."

"What would we hope to gain by disbanding?"

"If you surrender, sir, our other armies will give up too. Our disbanding, rather than surrendering, might ease its effect upon those who still can fight. And General," he adds, "taking to the hills is better than unconditional surrender, placing ourselves at their mercy."

"If we disband the army," Lee, forces Alexander to analyze his suggestion, "how many men do you suppose would get away?"

"Perhaps two-thirds of us, sir. We'd be like rabbits and partridges in this broken country. They'd not have enough regiments to find us all."

Silence, then Lee shakes his head. No.

"If we disband, our men will have no rations, no weapons, no military organization. Already demoralized, they'll have to steal to survive. Not a good ending for men who've soldiered so well. And not good for our country. No; perhaps some of our soldiers will turn to bushwacking, but the proper course for all of us is for me to surrender the army and take the consequences."

Longstreet, knowing Lee so well, reminds him, "The Sam Grant we know won't insist on unconditional surrender, General. He'll offer the best terms he can. Just as you would if the situation were reversed."

"Yes," Lee seems to feel better for the moment. "Gentlemen, please return to your commands. I must meet with General Grant beyond General Longstreet's lines at 10 o'clock."

Lee, accompanied by Confederate Colonels Walter Taylor and Charles Marshall and Sergeant George W. Tucker, General A. P. Hill's favorite courier who'd attached himself to Lee's headquarters after General Hill's death, nears Longstreet's picket line. Tucker carries a flag of truce, a dirty handkerchief tied to a stick.

To the west, behind them, the firing, which for a time was heavy, has sputtered then stopped altogether.

As the little group pass, Longstreet's men cheer Lee, just as they've always done, until he waves them to silence lest their cheering draw enemy fire.

At the picket line, Lee and Taylor stop while Sergeant Tucker and Colonel Marshall ride forward to meet the lone Federal officer riding toward them.

Lee expects the man to guide them to Grant. Such is not the case.

"Colonel Charles Whittier," the Federal officer introduces himelf to Marshall. "General Humphreys' staff. I have a message from General Grant for General Lee."

"But we anticipated meeting General Grant here," a confused Marshall protests.

"Sorry, Colonel," an equally puzzled Whittier shakes his head. "General Grant's not here, and all I know is that I'm to deliver this message. I'll wait for a reply."

Marshall takes the envelope to Lee. It's Grant's reply, written early that morning from Curdsville.

April 9, 1865

General: Your note of yesterday is received. As I have no authority to treat on the subject of peace, the meeting proposed for 10 a.m. to-day could lead to no good. I will state, however, General, that I am equally anxious for peace with yourself, and the whole North entertains the same feeling. The terms upon which peace can be had are well understood. By the South laying down their arms they will hasten that most desirable event, save thousands of human lives, and hundreds of millions of property not yet destroyed. Sincerely hoping that all our difficulties may be settled, without the loss of another life, I subscribe myself, etc.,

U. S. Grant
Lieutenant-General.

Lee, dismayed by the letter and puzzled by its contents, re-reads Grant's answer. It suggests a much broader discussion than he'd intended. Could Grant have misunderstood his intent? Worse yet, the phrase "By the South laying down their arms" leaps out at him. "Laying down their arms," the spectre of unconditional surrender raised again. The assurance he'd hoped for, that his surrendered officers and men will be treated as prisoners of war rather than as criminals charged with treason, is not there.

He reads the message again, stands silently thinking. Then he turns to Marshall.

"Well, Colonel Marshall, we have no choice. Take your pad. We must write General Grant again, asking him to meet me to arrange the surrender of my army."

As Lee begins to dictate his answer to Grant's message, he's interrupted by Confederate Colonel Jack Haskell, who's come from Longstreet. Haskell's horse, a large, beautiful stallion, is foam flecked from a hard run.

Lee's response is instinctive.

"What is it, Colonel? What is it? I'm afraid that you've killed your beautiful horse."

"Fitz Lee has found a way out, General," a chastened Haskell answers. "If we hurry, General Longstreet thinks that the army may be able to get away."

Lee briefly questions Haskell then, dismissing the notion as if he'd never heard it, finishes his dictation:

...I now request an interview in accordance with the offer contained in your letter of yesterday for that purpose."

As Whittier gallops away with Lee's reply, Lee turns again to Haskell.

"Colonel, I neglected to tell General Gordon of my planned meeting with General Grant. Please see that he is informed and have General Gordon arrange a truce with the enemy before him."

Then, his decision made, he takes a moment for another concern, calling to Haskell as the Colonel mounts his horse, "And Colonel Haskell, please don't ride your horse so hard."

GRANT

Porter had found him pacing the yard before his farmhouse head-quarters, long before daybreak but so tormented by the blinding headache that he's given up on rest.

They urge him to ride an ambulance today, like Meade. It would be easier than the jolting ride Cincinnati will give him. Grant, however, waves off the suggestion. With Sheridan at Appomattox Station, and Ord and Griffin hoping to be there at dawn, *that* will be the point of action. If another fight is necessary, he should be there too.

Ignoring Rawlins' protests, he leads them on a jolting cross-country ride along little-used trails, fording creeks and muddy bogs. They see more and more Confederate stragglers, bypassed by the thousands of Federal infantrymen who already have passed this way. Sometimes Grant points them out to his staff, but as usual he says little.

As the sun breaks through the early morning mists, somewhere ahead, as if drawing Grant forward, they hear the dull thump of artillery fire: slow, sporadic, then heavy, rhythmic, sustained. It eases, picks up again, then dies. Cincinnati, always hard to keep up with on a march, rides without Grant's urging, as if as determined as his master to get to the source of that sound as quickly as possible.

At last Grant glances up at the sun, now high overhead, studies his watch as if verifying its accuracy, and gives in to his headache. Leading them into a field being cleared of trees and underbrush, he dismounts to rest.

They are enjoying their cigars before a fire when a scout warns, "Rider, General. Behind us. Comin' fast. Wearin' a blue uniform."

The lone rider following their trail is moving fast, his large black stallion foam flecked from its hard ride. Then, seeing Grant's party, the rider waves his hat and shouts for their attention.

In a few moments Lieutenant Charles E. Pease of Meade's staff has reined his horse, saluted, and handed Rawlins a sealed envelope.

Rawlins tears it open, taking, it seems to the others, an eternity to draw out its message. Then he slowly, deliberately reads it before handing the single sheet to Grant.

Grant's reading seems as slow as Rawlins'. The message, it seems to his anxious staff, bringing no more expression to Grant's face than would "a last year's bird's nest."

Then Grant hands it back to Rawlins with the quiet suggestion, "You'd better read it aloud, General."

Rawlins, in a voice which trembles with emotion, begins.

April 9th, 1865

General: I received your note this morning on the picket line whither I had come to meet you and ascertain definitely what terms were embraced in your proposal of yesterday with reference to the surrender of this army. I now request an interview, in accordance with the offer contained in your letter of yesterday, for that purpose.

Very respectfully, Your obt servt

R. E. Lee

When Rawlins finishes reading the message, no one speaks. It's a moment that each of them will always remember. Its implications as broad as each man's own heart and dreams.

A younger officer springs upon a log and, waving his hat, calls out, "Three cheers, boys! Let's have three cheers for peace. Then three more for General Ulysses S. Grant."

They try to follow his lead, but it's not a success. Several of them muster the first feeble hurrahs then stop. The others, their throats too tight, their hearts too big for their chests, can't cheer. Several of them, overcome with emotion, openly sob. No one is ashamed of that.

They look at one another as men may do when they've been under an earth-shaking, deafening artillery barrage that suddenly ends, and they realize that they're still alive. Look into each other's hearts for something that only those who've been there can understand.

When the moment passes, as all such moments must perish quickly, they begin to cheer and to congratulate each other. Their little celebration beside a little patch of Virginia woods continues for some moments.

Grant wryly grins around his cigar and, recalling Rawlins' tirade a few hours before, asks, "How'll *that* do, Rawlins?"

"I think, General," Rawlins is firm in his answer, "*that* will do."

Warming his hands by the fire as he talks around his cigar, Grant dictates an answer to Lee's message to his military secretary, Colonel Ely Parker.

Parker, the most striking figure in Grant's party, is a full-blooded Iroquois Indian. Tall, round-faced, taciturn, Parker is a law school graduate who, finding that because of his race he couldn't successfully practice law, became the chief engineer of the Chesapeake and Albemarle Canal. Then, when the war began, Parker, a personal friend who'd helped Grant during the troubled days when Grant worked as a clerk in his brother's tannery in

Galena, Illinois, resigned that post to become Grant's personal secretary. He's known for his penmanship; he's also known among Grant's staff for an incident that occurred some months ago.

A man, claiming to be an old acquaintance of Grant's, had come to the army's headquarters to see Grant.

"Where's the Old Man?" he'd demanded.

Rawlins had been too busy to do more than point toward Grant's tent. The visitor, following Rawlins' gesture, saw Ely Parker framed in the tent's opening, talking to a hidden Grant.

"Yep," the man confirmed. "That's Grant all right. But he's got all-fired sunburnt since I saw him last."

When Parker has finished writing, Grant signs the message and hands it to an aide.

"Get it to General Lee as soon as possible, Colonel Babcock. Wait for a reply. You'll find us along this road or with Sheridan at Appomattox."

Then he turns to his party. "Let's be on our way, gentlemen. I want to get up to Sheridan. The firing up there stopped some time ago. Perhaps the issue's already been decided; if not, we may be able to end it without more bloodshed."

"How's your headache, General?" Porter asks.

Grant, puzzled for a moment, grins again, "Gone, Porter. Gone. Gone the minute I read that message."

LEE

Union Colonel Charles Whittier soon returns to speak with Colonel Marshall, waiting near Humphreys' lines.

"Colonel Marshall, General Humphreys asked me to tell you that Meade's ordered him to attack. General Humphreys understands your situation but says that only General Meade or General Grant can change those instructions. And General Grant's off with Sheridan somewhere ahead of us. He's afraid that General Lee's letter can't reach Grant in time to stop the attack."

"Ask General Meade to read General Lee's letter and to act upon it," Marshall argues. "Surely under the circumstances he'll delay his attack before useless blood is shed."

"General Humphreys' gone to see him, sir; but he suggests that you try in any way you can to get word to General Grant yourself." With that Whittier gallops away.

Marshall explains the situation to Lee, and Lee sends another message to Grant. This one, similar in meaning to his earlier message, is to pass through Ord's lines west of the village in hopes that it may reach Grant more quickly:

General,
I ask a suspension of hostilities pending the adjustment of the terms
of the surrender of this army, in the interview requested in my former
communication today.

Thinking that Whittier has been gone a long time, Marshall and Tucker
wait between the lines for Meade's answer. Meanwhile, Federal pickets
begin edging closer. Lee, quietly waiting on Traveller, sees them but obviously does not intend to withdraw. Seeing Lee's determination, Longstreet
moves his own infantrymen forward, should the Northern soldiers attack.

Humphreys' skirmishers are within a hundred yards of Lee before
they stop, stack arms, wait. Only then does Lee withdraw.

In a small apple orchard beside the Lynchburg Stage Road, where
he'll wait for Grant's reply, Lee receives two more messages. The first, the
report of an escape route for the army has been a mistake. Federal infantry
solidly block the road. Lee, not surprised, sends another message telling
Gordon to hold where he is.

A second messenger brings Meade's answer. The Union commander
will honor a truce for one hour, no more. Meanwhile, he also suggests that
Lee send another message to Grant. Lee dictates a fresh dispatch, ending:

I therefore request an interview at such time and place as you may
designate, to discuss the terms of the surrender of this army....

Colonel Talcott has posted guards around the orchard to give Lee
some privacy; and Porter Alexander spreads blankets across several fence
rails beneath a large apple tree and persuades Lee to rest there until he
hears from Grant.

A bit later Longstreet joins them there, and the three generals talk
quietly, Lee sharing his fear that Grant may demand harsh terms after all.

"No," Longstreet reassures him, "I know Grant. The terms will be
about what you would demand under the circumstances. No worse."

Then, seeing a Federal officer being escorted to the orchard, he growls
a last bit of support, "Look, General; if Grant won't give you honorable
terms, break off the talk. Tell him to do his worst, and we'll oblige him."

Lee straightens and smiles. Longstreet, his "old war horse," has come
through again.

"Bring the courier here, Colonel Marshall. Let's see what General
Grant has said."

The Federal courier, Colonel Orville Babcock of Grant's staff, salutes
then hands Grant's message to Lee.

General R. E. Lee

Commanding C. S. Army

Your note of this date is but this moment (11:50 A. M.) received. In consequence of my having passed from the Richmond and Lynchburg road to the Farmville and Lynchburg road. I am writing this about four miles west of Walker's church and will push forward to the front for the purpose of meeting with you. Notice sent on this road where you wish the interview to take place will reach me.

Very respectfully, your obedient servant,

U. S. Grant

Lieutenant-General

Lee asks Colonel Walter Taylor to go with him to meet the Union commander.

The youthful Taylor, dreading the meeting to arrange the surrender of the Southern army, begs off.

"Sir, I've ridden through the lines twice this morning, and I don't know that I can handle this meeting. Might I be excused?"

"Yes," Lee smiles gently. "Colonel Marshall, you will come with me."

Marshall, mindful of the magnificent appearance of his commander, pauses long enough to borrow a fresh shirt and collar he miraculously finds somewhere among the staff.

As he waits, Lee hands Colonel Venable the worn map of photographic linen he's used for months.

"Here, Colonel Venable. Burn this map."

Before Venable can destroy the map, Porter Alexander cuts from it a small strip bearing the notation: "South Side James River, R. E. Lee."

Lee, Marshall, and Private Joshua Johns, 39th Virginia Cavalry Battalion, accompanied by Federal Colonel Orville Babcock, ride up the long hill toward Appomattox Court House. The Confederate soldiers they pass recognize Lee and begin to cheer. When they see the white flag carried by Private Johns, however, their cheers die. Word of the flag quickly spreads. Then most of them simply stand bareheaded, silent, as Lee, looking straight ahead, passes.

Near the village Marshall and Johns go ahead to find a suitable place for the meeting. Marshall asks a local man, Wilmer McLean, for his suggestion.

Four years before McLean owned a farm on the site of the first major battlefield of the war: Bull Run to the Northerners, Manassas to the

Southerners. McLean's home and fields lay between the two armies and were ruined by cannon and rifle fire, by charge and countercharge. When the battle ended and the armies went away, McLean surveyed his ruined lands.

He'd moved his family to distant Appomattox Court House, where he hoped that he'd never again see either army nor hear another shot fired. Now he finds it a striking irony that, nearly four years later, he's asked by both armies to provide a place where they might end their war.

He suggests a vacant brick building, the former Clover Hill Tavern. When Marshall rejects the tavern's disheveled emptiness, however, McLean offers his own home, a red-brick, two-story dwelling with a broad, inviting porch, less than a block from the courthouse itself.

TRUCE ON THE FIELD

Earlier, beyond the village, Gordon's troops still were slowly, stubbornly withdrawing when Gordon learns that Lee is trying to arrange a meeting with Grant and wishes Gordon to arrange a truce with the Federal forces before him.

"Colonel," he turns to his staff, "get a flag of truce and go through the lines. Find General Ord. Tell him that General Lee has asked me to request a truce."

The officer turns to run to his horse then, puzzled, hesitates.

"We have no flag of truce, General."

"Well, make one. Tie your handkerchief on a stick."

"I have no handkerchief, sir."

"Then tear your shirt, Colonel, and tie that to a stick."

"General, I'm wearing a flannel shirt, like your own. I don't believe that there's a white shirt in the army."

"Well," Gordon smiles grimly, "find something, sir. Find something and go."

The Confederate riders with their white flags approach Chamberlain's and Custer's lines.

The first Southern officer, Major Hunter Jones, holding high a large, white napkin that he'd remembered having in his saddlebag, the wrapping for a piece of pork he'd been given the day before, salutes Chamberlain.

"Sir, I'm from General Gordon. General Lee asks that the fighting be stopped until he can hear from General Grant about the surrender."

"Surrender?" Chamberlain is stunned. He knew that the end must be near, but it didn't seem possible that it could come so suddenly.

"Surrender, sir," the Confederate officer softly answers.

"Well," the professor turned soldier takes a moment to compose his thoughts, then, "Surrender. Well, that exceeds my authority, but I'll send word to General Griffin."

Then, feeling a need to somehow comfort the Southern officer, he adds, "I understand your pain, but General Lee is right, sir. He really can do no more, has no other choice. Please wait now while I send for General Griffin."

Off to his right, Chamberlain watches as another Confederate courier, bearing a second small, white flag, approaches Custer's lines of cavalry.

Colonel Edward W. Whitaker, Custer's Chief of Staff, returns the officer's salute.

"Colonel, I'm Captain Robert Sims of General Longstreet's staff, attached to General Gordon. Where is General Sheridan? I have a message for him."

"He's not here, Captain; but General Custer is, and you'd better see him."

Custer's response to Sims' salute is brusque.

"Who are you? What do you want?"

"Captain Sims; General Longstreet's staff, General. I have a message from General Gordon to General Sheridan asking for a cease fire until General Lee can be heard from. He's gone down the road to meet General Grant."

"You can tell your general," Custer shakes his head and snaps, "that we'll listen to no terms but unconditional surrender. We're behind your army, Captain, and it's at our mercy."

Sims' face hardens. Then answers, "Is that the message you want me to take back to General Gordon?"

"Yes, go on."

Sims starts to rein his horse then turns again, "General, do you want to send an officer with me?"

Custer nods to Whitaker and another officer. As the trio gallop forward, Whitaker shouts to the line of cavalrymen, still poised to charge, "Put down your weapons, men. This is unconditional surrender! This is the end!"

They hear Whitaker's shout, and the word instantly spreads from regiment to regiment. Across Custer's and Chamberlain's lines Union soldiers immediately begin to celebrate. A dozen bugles sound and drummer boys, dozens of them, beat a steady tattoo. Soldiers shout, cheer, throw their caps in the air, hug one another, openly cry.

Their celebration will go on for half an hour until they're too emotionally exhausted to do more than simply wait; wait for the next step to occur.

Gray-haired Brigadier General Edgar M. Gregory, commanding a brigade in reserve and alarmed by the shouting among Chamberlain's soldiers, gallops up.

"What's happening, General? Why have we stopped? Why are the men behaving this way?"

Chamberlain grins, "Well, General, it seems that Lee wants time to surrender."

"Glory to God, Chamberlain!" the older general exclaims, reaching over to hug the Maine commander.

"Amen to that, General," Chamberlain grins again, "and on earth peace and good will towards men."

General Ord arrives, shouting to the soldiers as he gallops toward Chamberlain, "It's over, men! It's over! Your legs have done it."

Yes, Chamberlain nods, their legs have done it. Their legs and blood still fresh at Sayler's Creek, High Bridge, Farmville, Five Forks, the White Oak Road, and the Quaker Road. Their blood, no longer fresh but left forever at Cold Harbor, the Wilderness, Gettysburg, Antietam, Fredericksburg, and on a hundred other never-to-be-forgotten fields.

Then Ord grasps Chamberlain's hand, "Hold your regiments where they are, General. Remain alert, but wait for further orders." He begins to gallop away, then stops and turns to add, "You've done well, Chamberlain. I'll not forget."

As they ride toward Gordon's lines, Custer's emissary, Colonel Edward W. Whitaker, demands Sims' surrender flag.

Sims angrily answers, "I'll see you in Hell first, Colonel! It's humiliating enough to have to carry this white flag and to wave it for all the world to see, but you can bet your boots that I'll not let you have it as a monument to our defeat."

Meanwhile, Custer, impetuous and impatient, not waiting for orders, has borrowed a not quite white handkerchief and, waving it up and down, also rides toward Gordon's lines.

As he nears the Confederate regiments, he demands, "Who's in command here?"

"General Gordon."

Every eye in the 31st Georgia Infantry Regiment is on Custer: slender, graceful, his blonde hair down over his shoulders; his uniform a dark blue with immense shoulder straps bearing the silver stars of a general. Adorning his neck, a blazing red cravat with a gold pin of two-inch letters:

"CUSTER". Custer, riding a horse difficult to handle but doing it with ease. A superb horseman, coming with the air of a conqueror. Now within easy range.

A Confederate soldier, unimpressed, raises his rifle, muttering, "Wal, I reckon I'll get that one anyway!"

Before he can fire, however, another Southern rifleman knocks the rifle aside.

"Don't do it, John. If we've already surrendered, and you kill him, it'll cause trouble."

"That's not a white flag. I'm not bound to respect that rag," the man argues and is lifting his rifle again when the rest of his squad stop him.

Custer, unaware of his close brush with death, reins his horse before John Gordon, whips his saber in salute, and speaks with all the authority he can muster.

"I'm General Custer, and I have a message from General Sheridan. He asked me to present his compliments to you, General, and to demand the immediate and unconditional surrender of all your troops."

Gordon, little impressed, makes him wait a moment then slowly, deliberately drawls his answer.

"Well, General Custer, please go back and return my compliments to General Sheridan. But tell him that I shall not surrender my command."

Custer, red-faced, shakes his head and tries again.

"I don't think that you understand, sir. We have you surrounded. If there's any delay in your surrendering, we can annihilate your command in an hour."

"Well, General," Gordon again is deliberately nonchalant, "I reckon that I'm as aware of my situation as you or General Sheridan. I *won't* surrender, sir, and if General Sheridan decides to attack in the face of a flag of truce, the responsibility for the blood that's shed will be his, not mine."

Custer, angry but uncertain now, salutes again then gallops away, toward the village.

SHERIDAN

Soon after Custer has gone, another Federal officer, his orderly waving a white flag, also approaches Gordon's lines. Behind them ride a score of Blue cavalrymen led by Major General Philip H. Sheridan. Sheridan, mounted on an enormous, black stallion.

Watching the approaching riders, one of Gordon's sharpshooters raises his rifle.

"No," this time it's Gordon himself who intervenes, "you can't fire on a flag of truce."

The man lowers his rifle, considers a moment, then begins to raise it again. Seeing the motion, Gordon seizes the rifle's barrel and speaks more forcefully, "No, you must not shoot."

"Well, General," the marksman gives in, "let him stay on his own side of the line."

Recognizing Sheridan, Gordon rides forward to meet him.

"I'm Sheridan, General. I expect you're Gordon."

"General Gordon, sir; General Gordon."

"We have an unusual situation, Gordon," Sheridan ignores the correction. "General Grant has not authorized this suspension of the fighting, but General Ord has decided to honor a truce until Lee and Grant can meet."

Gordon nods his understanding but, before he can answer, a sudden burst of firing erupts from behind him.

"What's that firing, sir?" Sheridan demands. "We'll not stand still while your men continue fighting."

"It's my fault, General," Gordon answers. "General Gary's brigade. I forgot to notify them. I'll do so now."

"Oh, let them fight it out," Sheridan retorts.

"No," Gordon answers, "enough lives have been lost."

As Gordon has no other staff officers available, he sends two Union officers, escorted by a Confederate captain, to spread the word that the fighting has ended.

Gordon had asked Captain Sims to escort the Union officers but when Sims, deciding that he must return to Longstreet's headquarters, declined, Gordon turned to Captain John Badger Brown to escort Sheridan's officers. Sims allows Brown to use his white "flag" but takes a moment to warn him, "Don't let that damn Yankee Colonel (Edward Whitaker) get his hands on it."

Somehow, however, Whitaker *does* get Sims' flag and later presents it to Elizabeth Custer.

When Gordon and Sheridan return to their conversation, Sheridan remains belligerent, arrogant.

"We've met before, General Gordon. At Winchester and Cedar Creek in the Valley," Sheridan boasts. "Matter of fact, I received some artillery there, consigned to me through your commander, General Early."

"Yes," Gordon drawls, "that's correct. And this morning you returned the favor when I received from your government artillery consigned to me through General Sheridan."

"You are mistaken, sir!" Sheridan's face is fiery red. "I've lost no cannon today."

"Perhaps," Gordon again is insultingly slow in drawling out his answer, "but if you'll look over there by the courthouse you'll see the two cannon, their accoutrements, and their teams. They have your unit markings, and somewhere back there are about a dozen of your artillerymen."

Sheridan, deciding that more conversation with Gordon won't gain much, rides into the village to see for himself.

Along the way, with Gordon in mind, he snarls to a staff officer, "Damn them! I wish they'd held out one hour longer. I would have whipped the hell out of them!"

During the armies' stay at Appomattox, Gordon and Sheridan will have little more to say to each other.

LONGSTREET

After leaving Gordon, Custer, accompanied by a Confederate officer and waving his white handkerchief, reins before General James Longstreet. He'd asked for Lee but had been directed to Longstreet instead.

"General Longstreet," Custer calls out loudly enough to be heard by all of Longstreet's staff, "I am General Custer. In the name of General Sheridan and myself, I demand the surrender of this army."

Longstreet takes several thoughtful puffs on his pipe then casually returns Custer's salute before he answers.

"I'm not in command of this army, General; General Lee is. You'll have to talk with him about that, and he's gone to talk with General Grant."

"Well, no matter about Grant," Custer gestures rapidly with his hands. "We are here, and we demand that the surrender be made to us. If you don't, we'll attack; and any blood shed will be upon your head."

"Young man," Longstreet is angry now, his voice contemptuous, "you've come into my lines under a flag of truce, without my permission. You've demanded that I surrender my command; something that I have no authority to do. And you, sir, are speaking disrespectfully to a senior officer. I suppose, General, that you don't know any better so I'll excuse you this time, but that won't save you if you do it again."

"Now, seeing that your blood is up," he continues, "let me accommodate you." Gesturing with his pipe, he orders, "Colonel Manning, have General Johnson move his division on line to General Gordon's right. Colonel Latrobe, have General Pickett move forward to General Gordon's left. Do it now!"

Then, turning again to Custer, "Now, General, I suggest that you and Sheridan do as you wish; and I will teach both of you a lesson you won't forget."

Custer again backs away, "General, you misunderstood. We *should* wait until we hear from Grant and Lee. I'll speak to General Sheridan about it. Don't move your troops yet."

When he's gone, Longstreet laughs, "Ha! That young man could never play Brag!"

They all know that after Sayler's Creek Johnson's and Pickett's divisions no longer exist.

Soon after the truce began, Gary's South Carolina brigade, not knowing that one was in effect, charged the Union lines before them, driving them back. This the fighting Sheridan and Gordon heard. Gary, with drawn sword, is preparing to continue his attack when a Federal officer, escorted by a Confederate captain and waving a white flag, arrives with Gordon's message.

The Union officer, Lieutenant Vanderbilt Allen of Sheridan's staff, demands, "General, why are you keeping up the fight after your army has surrendered?"

"Surrendered," the fiery Gary shouts. "I've heard of no surrender. We are South Carolinians, sir, and South Carolinians do not surrender. Besides, I take commands from no officers except my own."

Allen, conciliatory now, explains, "General, I appreciate your feelings, but there's been a truce pending negotiations. I've been sent to tell you that Generals Grant and Lee are conferring now."

Gary explodes again. "Do you expect me to believe such damned talk as that? Where's the officer who brought you here? Meanwhile, you, sir, shall go to the rear as a prisoner."

Allen's Confederate escort salutes, "Captain Blackford, sir. I'm afraid that the report is true, General. I can verify it."

For a moment Blackford is not sure whether he or the Yankee officer will be cut down by Gary's sword. Then the red-faced general, looking very old and very tired, returns the blade to its scabbard and, lowering his voice, turns again to Allen.

"Captain, I don't question your truthfulness, but you know that I can take orders only from my own officers. I'll accept your report, however, call back my men, and await those orders. Captain Blackford, please report the circumstances to General Gordon. We'll take no further action until I hear from him."

Later, when Gordon confirms the pending surrender, Gary and several other officers will quietly slip through the Federal lines to make their way to Johnston's Confederate army in North Carolina.

Word of the truce doesn't reach everyone at the same time, and that causes more confusion and uncertainty. Soon after the white flags appear, soldiers of the 11th Maine Infantry Regiment are startled by sudden cheers and a burst of firing from Ord's regiments on their left. Then they learn that it's not an enemy attack but a brigade of Black soldiers of Birney's division, celebrating the truce the Maine men haven't yet heard about.

In Union Brigadier General Edward Gregory's V Corps' reserve brigade the Chaplain of the 189th New York Infantry Regiment is leading a prayer of thanks when they hear the heavy firing involving Gary's South Carolinians. The chaplain quietly cuts short the prayer, explaining, "Never mind, boys; we'll come back to it later. Form your battle lines."

When the firing ends, the soldiers quietly gather around their chaplain again, and he picks up the interrupted prayer.

In the Confederate lines any soldier who dares suggest "Surrender" is apt to be beaten by his comrades.

Union Brigadier General Gregg, captured in the fighting at Cumberland Church, is among the Union prisoners alerted by Confederate guards that they soon will be sent back to the Union lines. First, however, Southern infantrymen will need the prisoners' hats and boots for themselves.

Most of the Northern prisoners, glad to be rejoining their own army, don't object to the trade. Gregg, however, refuses to give up his fine boots. Several Confederate soldiers hold his arms while two more wrestle the boots from his feet.

Like the fine hat that he'd lost to a Georgia private two days before, Gregg can only helplessly watch the boots' new owner try them on then, grinning, wave Gregg an impudent salute and walk away with them.

The Color Sergeant of the 8th Alabama Infantry Regiment holds his bullet-torn regimental flag in his hands, nervously trying to straighten wrinkles in the torn banner as he considers all the flag has seen, the battles it's known.

Then, his decision made, he speaks to the flag as if it were a living person, "You've never run in battle; and you don't surrender!" Tearing the flag into little pieces, he parcels it out among the men.

Another Confederate veteran, Sergeant W. E. Deas, puffs his pipe and accepting the inevitable, shares, "Ah, a wretched day. And right now it seems that the Almighty has hidden his face from us. But I trust in Him yet and know that He does all things well."

Advancing Federal skirmishers stop in the village, at the crest overlooking the valley of the Appomattox, stack their rifles, and cautiously wait. Several hundred yards down the slope before them the slowly withdrawing Confederates stop too, stack their own weapons, and wait.

As soon as the truce is confirmed, however, Confederate soldiers, unarmed, immediately move to cross the field to their enemies' lines. They're received kindly, and soon Blue and Gray soldiers are busy trading knives, harmonicas, and other trinkets. Already they're sharing food from Federal knapsacks.

Chamberlain, riding Charlemagne slowly into the village, sees a large cluster of Federal and Confederate officers talking before the brick courthouse at the end of its lone street. He stops short, however, of joining them.

Turning in his saddle, he thoughtfully scans the wreckage about him. Torn and splintered trees, fence rails shot to pieces, and dead men and horses from last night's fight are everywhere. Union and Confederate dead lie in the yards, beside the road, against the fence lines. Some with their eyes wide open, still grasping their rifles. A Union cavalry colonel, stripped of his boots, lies dead in the road. Wounded men, mangled by cannon fire, wait for someone to come help them. A Confederate cannon still is in position, once a deadly presence but now harmless without its crew, on a little, rutted dirt road in a village none of them had ever known existed, at the very end of the war.

Chamberlain senses a new presence. From the east, up the stage road, three Confederates and a lone Union officer ride slowly toward him.

He can't mistake the central figure: General Robert E. Lee, tall, erect, disciplined, superbly dressed and superbly mounted. The big gray horse he rides has been brushed until his coat shines and the metal of his harness polished until it glints in the early afternoon sun. Traveller, not understanding the purpose of this trip but knowing from the attention he's received and the appearance of his master that it must be very important, tosses his head and mane at the sight of Charlemagne and breaks into a short canter.

Horse and man: beautiful; one. Lee: grave, sad, projecting deep strength. Chamberlain, as far now as he'll ever be from his quiet Maine campus, sits erect in his saddle but, overcome by the moment, forgets to salute Lee as the Southern general passes.

Lee seems not to notice Chamberlain's oversight, thoughtfully smiling at him then passing on to turn into the yard of a large, brick home.

Private Johns dismounts to take Traveller's bridle as Lee dismounts.

Then Lee, Marshall, and Union Colonel Babcock enter the McLean house.

There is a parlor, a small room to the left of the center hallway. Lee chooses a chair there, near a small table at the front of the house where he can look across the valley of the Appomattox. He places his hat, sword, and gauntlets on the table, and crosses his legs, a disciplined relaxation he does not feel. His new boots are embroidered in red silk. The handsome sword glitters in its scabbard. One side of its blade is engraved: "General Robert E. Lee, from a Marylander, 1863; the reverse edge is engraved: "Aide toi et Dieu t'aidera," "Help Yourself and God Will Help You."

About a half hour after Lee has passed, approaching from the opposite direction, Chamberlain sees Grant. Ulysses Simpson Grant, with a single aide to guide him to the meeting where he's promised, "Lee's a good man. If I can sit down with him for an hour, we can end this thing."

Grant: slouched in his saddle; wearing a worn, weather-faded hat without a cord and the uniform of a common soldier with the stars of a Lieutenant General tacked to its shoulders. Grant's blouse is unbuttoned. His trousers are jammed inside his boots like an Ohio farmer's; trousers and boots that are mud-spattered. He wears no sword. His sword hand is in his pocket while he casually allows Cincinnati to pick his own way. A dead cigar is in his mouth.

Grant, recognizing Chamberlain, smiles around his cigar as he returns Chamberlain's salute.

Grant's eyes, Chamberlain realizes, are as drawn and as tired as his own. Eyes with that thousand-yard stare combat soldiers share.

Ulysses Simpson Grant: no pomp and circumstance, not very imposing in appearance, but somehow, like Lee, projecting great strength, patience, and endurance.

"Good morning, Chamberlain. How are you?" His greeting as casual as if they were meeting in another village, a thousand miles away, in another time.

"Fine, General."

Sheridan and Ord are there to meet Grant.

When Sheridan salutes, Grant smiles, "How are you, Sheridan?" as casually as he'd greeted Chamberlain.

"First-rate, thank you. How are you?"

"Better than toler'ble, I'd say," Grant allows himself another grin around the cigar, "better than toler'ble."

Then, "Is Lee here?"

"Yes," Sheridan points to the McLean house. "He's there, in that brick house."

"Well, then, we'll go there."

Sheridan leads the way to the McLean yard. There Grant dismounts, takes a moment to gaze around the village, as if seeing it for the first time or perhaps to clear his mind.

Seeing Private Joshua Johns, Confederate States Army, standing beside Traveller, he smiles as one man will to another, then returns Johns' salute.

A beautiful horse, Grant decides. It doesn't occur to him at the time, though had someone else suggested it, it certainly would have interested him, that three of the most famous horses in American military history: Traveller, Cincinnati, and Rienzi, will graze together in that simple little yard.

Then he mounts the steps to its porch.

Inside, the three men have been quietly talking when they hear voices outside the house then heavy boots on its porch. Lee turns in his chair toward the doorway and his first glimpse of the man he'd last seen in Mexico, in the Old Army, in another war, in another time, in whose hands he now must place himself and his army, but whose features he just can't recall.

CHAPTER 8

Two Good Men   Palm Sunday, April 9, 1865

Wilmer McLean's home, easily the most impressive residence in the little village, sits back from the street, its entrance guarded by locust trees, a white picket fence, and a greening lawn. Across its front is a white-columned full porch. Seven wide steps lead from the yard to the porch.

Inside its front door a center hall runs from front to back, with a large room on either side of the hallway. Each room has two doors from the hallway, and two windows: one window facing the stage road before it; the other overlooking McLean's fields, sloping to the west and south.

Lee and Marshall are in the parlor to the left, to which Colonel Babcock escorts Grant. Lee is sitting in a cane-bottomed arm chair with a tall back, in a corner beyond the front window. To his left, near the fireplace, Colonel Marshall stands. Near Lee is a small marble-topped table, upon which he's placed his hat and gauntlets. In the center of the room is a small chair which will be used by Grant. Centered between the two parlor doors is a large sofa and, nearby, there is another small table which Grant later will use. The fireplace is softened by various knickknacks on its mantel; the walls adorned with several nondescript prints.

Grant, climbing the porch stairs, pauses to consider his dirty, wrinkled, muddy uniform but, realizing that there is nothing that he can do about it now, takes a deep breath and follows Babcock into the house.

As Grant enters the parlor, Lee, Marshall, and Babcock instantly rise to greet him. For just a second Lee almost smiles, a personal thing that comes with his now recalling Grant's features in Mexico so long ago. Marshall almost smiles too, but for a different reason. He's struck with the thought that by the commanders' appearances it is Lee who seems to be the conqueror, Grant the conquered.

Lee, a sturdy, vigorous fifty-eight years old, is nearly six feet in height and very erect. His silver-gray hair has receded slightly but, with his beard, he still is a handsome man. He's wearing the new Confederate uniform he'd donned that morning; its gray blouse buttoned at his throat. His boots, of rich leather with red silk stitching, are well polished and adorned with large-roweled spurs. A deep red sash and sword belt are about his waist; the sword an expensive and beautiful blade. All worn with simple, natural grace.

Grant is some sixteen years younger than Lee. His hair and short-cropped beard are a nut-brown with just a bit of gray beginning to appear. He's five feet eight inches tall with a slender build, but his slightly stoop-shouldered posture and tired, worn appearance make him appear older and shorter than he is.

A man of scrupulous personal cleanliness Grant, General-in-Chief of all the Northern armies, today is streaked with mud from his cross-country ride. He holds a nondescript, cordless, black slouch hat in his hand; has no sword or sash; wears the uniform of a common soldier but with the stars of a lieutenant general tacked to its shoulders, as if by an anxious aide fearful that his general might not look the part.

Colonel Horace Porter, also comparing the two men and perhaps having noted Marshall's initial impression of Grant, is reminded of the time an army drover, failing to recognize the ranking general in the United States Army, asked Grant to shoo a straying cow back into his herd; and Grant did it. And when Grant threw his cigar into the James after a soldier, not recognizing him, ordered it. And the time Grant pitched in to help Negro workers roll barrels of hardtack from a supply boat.

Porter, however, knows what the man has done and is capable of doing and probably is the least concerned about Grant's appearance.

"General Lee," Grant speaks first, removing a pair of worn, yellow gloves as he crosses the room, "I'm very glad to meet you."

"General Grant," Lee responds and firmly takes the extended hand.

Grant takes the chair in the center of the room while Lee returns to his, perhaps five feet nearer the window. Marshall stands by the mantel.

Babcock then quietly slips from the room to return with a dozen additional Federal officers: Sheridan, Rawlins, Porter, Parker, and Grant's aides: Captains Robert T. Lincoln and Adam Badeau, and others of Grant's personal staff. It's about 1:30 in the afternoon as they quietly take their places on or near the sofa to witness the historic meeting. If their presence troubles Lee, he does not show it.

Throughout their meeting Lee's disciplined face and demeanor reveal little of his feelings. Grant, on the other hand, on the greatest day his life has known to this point seems sad, preoccupied; almost as if he's the one who has lost. He's a modest, sensitive man who knows all about personal failure. So he sympathizes with the situation his rival, who has fought him long and well, now feels and, in his meetings with Lee, will be very sensitive to the Southern commander's feelings.

Grant opens their conversation, as most soldiers are inclined to do, by recalling the last duty station they'd shared.

"I met you once before, General, but I doubt that you'll remember me. We were serving in Mexico. You were General Scott's Chief of Staff and had come to visit my brigade, Garland's. I admired your appearance that day, and I've always remembered it. I think that I would have recognized you anywhere."

Somewhat to Grant's surprise and pleasure, Lee does recall their meeting.

"Yes, I recall meeting you that day, and I've often thought about it and tried to recollect how you looked. But I'm sorry to say that, until you entered this room, I couldn't recall your exact features."

Their conversation about their days with the Old Army relaxes Grant. He comments again on the memory he's always carried of Lee as a young captain, compliments him on his appearance now, and apologizes for his own dress. Lee gently smiles and dismisses Grant's concern.

"General, I'm grateful that you hurried to our meeting. That is all that matters."

For some moments Grant, as if hesitant to approach the question of the surrender of Lee's army, recalls other incidents that might have significance to the other man. Lee nods at them, smiles, occasionally adds a thought of his own.

Finally, it's Lee who steers him to the purpose of their meeting.

"General Grant," he begins, "we both understand that the purpose of our meeting is to determine the terms under which you would receive the surrender of my army."

Grant, drawn from the conversation he'd been enjoying and keenly aware of the humiliation the older man must feel, turns to the unpleasant business at hand. His tone is warm, friendly, reassuring, but firm.

"General, the terms I propose are substantially those I stated in my letter yesterday. That the officers and men who surrender will be paroled; free to return to their homes. Of course they would be disqualified from taking up arms again until they be properly exchanged for other prisoners.

And all the arms, ammunition, and supplies of your army must be delivered up as captured property."

Lee breathes a deep but almost imperceptible sigh of relief, his eyes never losing contact with Grant's. So it is not to be unconditional surrender; his officers and men would not be marched off in chains. No retribution. Thank God.

He nods, "Yes, those are the conditions I expected you would propose."

Then Grant suggests the possibility of a broader outcome.

"I hope that what we do here may lead to an ending to this war."

Lee nods, but brings him back to the matter at hand: the surrender of the Army of Northern Virginia.

"I'm sure that we've both carefully considered the steps we must take. May I suggest that you write the terms you have proposed. Then we can formally act upon them."

"Of course," Grant nods. "I'll write them out."

He glances at his secretary, Colonel Parker, who produces Grant's field order book, a memo pad in which several copies are carboned to the original. Parker brings the second, small, oval table to Grant's side and places the pad upon it.

Grant opens the order book, studies it a moment, finds a pencil. Thinking about what he will write, he instinctively begins to light a cigar then, realizing where he is, looks for Lee's approval. When Lee smiles and gestures with his hand, Grant lights the cigar and becomes totally immersed in what he is doing.

He knows the terms he wishes but, up to that moment, he's not considered how to formally express them. Once he begins to write, however, his work goes quickly. When he's composed several paragraphs he stops, glances thoughtfully at the sword at Lee's side, then continues writing.

Finally he's done. Then he re-reads his work, makes several corrections, and nods to Colonel Parker. When Parker has joined him at the small table, the two men talk quietly, Parker leaning over to make several more minor corrections in the text.

Then it's done, and Grant half rises to take the order book to Lee. Colonel Horace Porter, however, takes the pad from his hand, crosses the few feet to Lee's table, and places it there for the Gray commander.

Lee, anxious to see the precise wording of Grant's terms, maintains the discipline that betrays none of his feelings. He deliberately takes his time, setting aside two brass candlesticks, some books, and his hat and gauntlets to make space on the little table. When he's satisfied with the arrangement, he opens the order book to Grant's work.

Before he takes up the pad, however, he draws a pair of steel-rimmed spectacles from his blouse and slowly, carefully cleans their lenses with his handkerchief. Satisfied, he adjusts them to his nose, straightens his shoulders, crosses his legs, and at last gives his full attention to the two-page document.

Appomattox Court House, Va., April 9, 1865
General R. E. Lee, Commanding C. S. A.
General: In accordance with the substance of my letter to you of the 8th inst., I propose to receive the surrender of the Army of Northern Virginia on the following terms, to wit: Rolls of all the officers and men to be made in duplicate, one copy to be given to an officer to be designated by me, the other to be retained by such officer or officers as you may designate. The officers to be given their individual paroles not to take up arms against the Government of the United States until properly, and each company or regimental commander signs a like parole for the men of their commands. The arms, artillery and public property to be parked and stacked, and turned over to the officers appointed by me to receive them. This will not embrace the side-arms of the officers, nor their private horses or baggage. This done, each officer and man will be allowed to return to his home, not to be disturbed by the United States authority so long as they observe their paroles and the laws in force where they may reside.
Very respectfully,
U. S. Grant
Lieutenant-General

Reading slowly and carefully, Lee has reached the top of the second page of the memo pad before he raises his head.

"General, after the words 'until properly' the word 'exchanged' seems to have been omitted. You probably intended to use that word?"

"Why, yes," agrees Grant. "I thought I had put in the word 'exchanged'."

"I presumed that you'd inadvertently omitted it," Lee replies. "With your permission, I'll mark where it should be inserted."

"Certainly. Thank you."

Lee searches vainly in his pockets for a pencil until Porter comes to his rescue with one of his own. During the remainder of their meeting Lee will twirl the pencil on his fingers or softly tap its end against the table.

He inserts a caret and, above it, the word "exchanged." Then he resumes his careful reading. When he comes to the vital provisions that his officers may keep their side arms, horses, and baggage; and that officers and men may return to their homes, not to be disturbed so long as they observe their paroles and local laws, he sighs deeply and, for the first time, his features relax.

Touching the pad with his pencil, his face softened by a slight smile, and with warmth in his voice, he nods to Grant.

"This will have a very happy effect on my army."

Grant nods, takes another puff of his cigar, then suggests, "General, unless you have some other suggestion, I'll have the letter copied and sign it."

Lee hesitates. Then, with a voice that is hesitant, strained, tight, he ventures, "There is one more thing that I would like to mention." He hesitates again then continues, "The cavalrymen and the artillerists in our army own their own horses. We differ in that respect from the Army of the United States. Would these men be permitted to retain their horses?"

Grant glances up at Rawlins, both men noting Lee's referring to the Confederacy as a separate nation. This time it's Grant, however, whose face does not betray his feelings.

"Well, the terms I've written," he pauses then continues, "state that only the officers would be allowed to take their private property."

Lee, his face flushed, sighs, drops his eyes to re-read the second page as he regroups. Then he looks up at Grant.

"No, I see that the terms, as written, do not allow it. That is clear."

There is such regret, such disappointment, in Lee's voice that Grant can't mistake his mute plea.

"Well," he comes to Lee's rescue, thinking out the problem as he speaks, "I didn't know that any private soldiers owned their own animals."

"I think, General," he continues, "that the other armies, when they hear of the surrender of your army, will surrender too. If that happens, we may have fought the last battle of this war. I sincerely hope so. And I know that most of your soldiers are small farmers. They're going to need their horses to get in a crop to carry themselves and their families through the next winter."

He puffs on his cigar, studies the fields beyond the window, recalls Lincoln's advice to Sherman, "Let 'em up easy, General; let 'em up easy." Then, his mind made up, he continues.

"I won't change the terms I've written, but I'll instruct the officers who will receive the paroles to let any man who claims to own a horse or mule take the animal home to work his farm."

Lee's smile is broad, grateful. They all feel as if a great load has been taken from him, from all of them.

"This will have the best possible effect upon my men," Lee confesses. Then he reflects, "It will be very gratifying, and it will do much to reconcile our people."

Colonel Porter returns the order book to Grant who reads the letter once more then asks Colonel T. S. Bowers to copy it in ink.

Bowers, young, awestruck by what he is witnessing, confesses, "General, I'm sorry, but I'm too nervous to do it. Might Colonel Parker replace me?"

Grant and Lee smile at the younger officer's confession. Then Grant nods and hands the book to Parker who takes it and the small table to a far corner of the room where he can work.

Then Parker stops before he's really begun, glancing up at the doorway. "Mr. McLean, I have no ink. Would you have some I might use?"

McLean disappears to return with a stoneware inkstand. Parker, however, finds that its ink is congealed, useless.

"Here, Colonel; I have some," and from a small satchel Colonel Marshall produces an inkstand and pen.

As Parker works, Grant introduces the other Federal officers to Lee and Marshall.

Lee is courteous, formally shaking hands with those who offer theirs, bowing to those who do not. His expression doesn't change except when he's introduced to Parker and to General Seth Williams.

Perhaps astonished that Grant should have an Indian on his personal staff, Lee offers his hand first, remarking, "I'm glad to see one real American here."

Parker, taking his hand, answers, "We're all Americans here, sir."

Then Lee speaks warmly with General Seth Williams, Grant's Adjutant General. Before the war, when Lee was West Point's Superintendent, Williams had been his adjutant and a personal friend.

Grant and Lee talk quietly while Parker continues his work.

It's then that Lee brings to Grant's attention an immediate problem.

"I have a thousand or more of your men as prisoners, General Grant. They've marched with us and shared the little food we have. Parched corn mostly. As I have no food for them, I'd like to return them to your lines as soon as possible. Indeed, I have none for my own men."

"I'd like for the prisoners to be returned as quickly as possible," Grant nods, "and I'll try to supply your army with rations. How many men do you have?"

Lee raises his hands, palms up, in a gesture, "I'm not sure. Our casualties in killed and wounded have been very high and our commands badly scattered. There has been much straggling and men have deserted their ranks. I'm just not sure how many are left."

"If I can provide 25,000 rations, do you think that will be sufficient for today?"

"Yes, that will be more than ample." Then Lee adds, "And it will be a great relief."

"Colonel Morgan," Grant calls his commissary officer, "can we send 25,000 rations to General Lee's army?"

"Yes, sir," Morgan figures aloud. "We carried twelve days' supply when the campaign began, and this is the twelfth day. But we have a little from the captured Confederate trains, and additional rations may reach us tomorrow. And our men already are sharing what they have with General Lee's men. If we can issue everyone short rations today, by tomorrow we should be fine."

"All right. See to it as quickly as possible. Beef, salt, hardtack, coffee, sugar. Forage for the animals. Whatever we can spare."

As Morgan hurries from the room, Grant's eyes again fall on Lee's handsome sword, and he tries again to explain his own appearance.

"I wouldn't have you take my appearance as a sign of disrespect, General. You see, when I left my headquarters several days ago I neglected to carry a sword. Frankly, I usually don't carry a weapon. And, when your message reached me, I thought it more important that we meet as soon as possible, regardless of my appearance, to prevent more loss of lives."

"General Grant," Lee again eases his concern, "I am much obliged to you and grateful that you did it this way."

Among the Federal officers there, Sheridan alone remains belligerent, choosing the moment to call Lee's attention to the Confederate cavalry's firing at him that morning as he approached Lee's lines.

"I mistakenly sent both copies of my formal note of complaint to you, General Lee, and I've not yet heard from it. I'd like to have one copy returned for my records."

Unruffled, Lee smiles patiently and, reaching into his tunic, withdraws several pieces of paper.

"I'm sorry, General. I just received your letters. I fear that my cavalry, at that point, didn't fully understand the truce. I'll look into the matter as soon as I can and report to you."

Colonel Marshall eases the awkward moment with a draft reply for Lee's signature. Lee makes several minor corrections then nods approval:

Headquarters, Army of Northern Virginia

April 9, 1865

General: I have received your letter of this date, containing the terms of the surrender of the Army of Northern Virginia as proposed by you. As they are substantially the same as those expressed in

your letter of the 8th inst., they are accepted. I will proceed to designate the proper officers to carry the stipulations into effect.

Very respectfuly your obedient servant

R. E. Lee

General

When both documents have been copied, they are signed and exchanged by their respective generals. The meeting is over; its business
done. Adam Badeau, historian and aide to Grant, glances at his watch.
Their meeting, ending nearly four years' fighting between the two armies,
has taken less than two hours.

When Grant and Lee stand, then shake hands, the remaining officers
stand at Attention as Lee prepares to leave.

Outside the house, Private Johns, Lee's orderly, had loosed Traveller's
bridle, allowing the horse to join the other two Confederate horses grazing
in McLean's yard. Then, hearing chairs being pushed back inside the house,
the noise of boots on its wooden floor, and men's rising voices, he readies
the three horses for Lee's return.

Above him the door to the porch slowly opens. Then Lee, bareheaded,
stands in the open doorway. He pauses there, gazing across the open
fields above the village, toward his own army drifting back to the camp they
are establishing in the valley of the Appomattox.

The clear, ruddy complexion of Lee's face now is a deeper crimson
that reaches up from a face tanned from days in the open sun and wind
into the paler white where his forehead has been protected by a hat. A face
strong and handsome but now grave, sad, thoughtful. Lee, magnificent in
appearance but sad-faced and weary.

Nearby Federal officers salute as he crosses the porch. Lee places
his hat on his head then mechanically returns their salutes. He pauses
again at the head of the porch steps. All the while searching the valley
where his army waits.

He slowly draws on his gauntlets and, as if oblivious to the presence
of anyone else, several times thoughtfully strikes one gloved hand into the
other.

Then, recalled to the job still before him, he glances deliberately left
and right and, not seeing Private Johns or the horses, calls in a strained,
half-choked voice, "Orderly! Orderly!"

"Here, General, here!" the Gray soldier is there in a moment. He slips
the bridle over Traveller's head and waits for Lee to descend the steps.

As Johns is buckling Traveller's throat latch, Lee reaches up and, as from long habit, draws the horse's forelock from under the browband, parts and smooths it, then gently pats the gray charger's forehead. Lee's mind seems faraway, but he's responding instinctively as one who loves animals. As he pats Traveller again, the horse raises and lowers its head in answer.

When Johns steps aside, Lee takes the bridle's reins and the pommel of his saddle in his left hand and, with his right hand on the saddle's cantle and his foot in the stirrup, swings himself slowly, wearily, but firmly into the saddle. The old "Dragoon Mount," just as he'd been taught as a cadet at West Point thirty years before and practiced countless times since. Learned. Disciplined. But there is a slight slip in his armor; a long, low, deep sigh escaping his lips while the flush on his neck deepens.

Behind him Colonel Marshall, slight and erect in his own gray uniform and heavy spectacles, also has mounted.

Seeing that they are ready, Lee slowly turns Traveller from the yard.

As Lee passes, a late-arriving *New York Tribune* reporter, Thaddeus S. Seybold, points to Lee and asks, "Who's that distingished-looking officer?"

A nearby Confederate officer, without taking his eyes from Lee, responds through his tears, "That, sir, is the greatest man America has ever produced, General Robert E. Lee."

Grant, also leaving, pauses at the foot of the porch steps to allow Traveller to pass. He looks up into Lee's eyes then raises his hat in a final salute; Lee returning the courtesy before he passes from the yard onto the stage road. Lee turns right, toward the courthouse and his waiting army.

Grant thoughtfully watches Lee ride away then turns back to Sheridan.

"Where will you make your headquarters tonight?"

"Probably right here," Sheridan answers.

"All right. I'll send some additional word. General Lee asked that fraternization not be encouraged. I doubt that we can prevent it, but let's have a corporal's guard for that purpose."

When Sheridan nods, Grant concludes the business, much as one businessman might turn from another, "Very well, then. I'll know where to find you. Good day."

As Grant turns west toward Appomattox Station and his own headquarters, he hears cheers erupting behind him: Confederate soldiers honoring their commander as Traveller finds his way back to the little apple orchard below the village.

With Grant and Lee gone, pandemonium breaks out in Wilmer McLean's home. Witnesses want souvenirs of the historic event. Furnishings, bric-a-brac, wallpaper, anything that will attest to their role in the surrender.

Never mind that McLean doesn't want to sell a single thing, refuses the greenbacks thrust into his hands or jammed into his coat pocket, and at last angrily throws a wad of the money into his yard.

Never mind all that. Take it anyway. A shabby ending to the high plane the two commanders had established for their meeting. Northern officers not giving the matter much thought; they've won, and they'll do as they wish.

Sheridan forces a $20 gold piece into McLean's pocket. This for the table Grant used to sign the surrender document. Outside, he presents it to the waiting Custer, a present for Elizabeth Custer. Custer, shouting his triumph, swings down from his saddle to take the table in his arms. Then, holding it over his head like a captured banner, he gallops back to the adulation of his waiting regiments.

Ord pays $40 for the table Lee has used. He'll present it to Mrs. Grant, but she'll decline and suggest it go to Mrs. Ord. General Sharpe pays another $10 for the brass candlesticks, and another officer takes away the McLean inkstand. Other officers claim the chairs Grant and Lee occupied.

Younger officers, seeing nothing else worth taking, seize Lula McLean's rag doll, "Lula", and, as they carry it from the house, toss it back and forth, laughing and calling it "the silent witness." Perhaps it's just as well that neither the doll nor its little owner can answer them.

When the other major furnishings have been taken from the room, several remaining officers, determined to own a share in the historic event, tear long strips from the horsehair sofa and from cane-bottomed chairs, breaking them into little pieces that the souvenirs may go further.

A tawdry ending Lee would have felt not worth his response but which an embarrassed Grant probably would have punished. Neither will know how the affair is conducted and perhaps that is just as well.

As Grant rides west, Lee east, sentries are posted to maintain order between the armies. Officers and men of both armies, however, already have begun to mingle, greeting each other cautiously at first then as friendly as if they'd never parted.

Before leaving the McLean house Grant sent word of the surrender to Meade. He'd realized, too late, that Meade's not being notified in time to attend the meeting will have repercussions. Meade, already feeling over-shadowed by the lesser ranking Sheridan, will take it as a personal affront.

"Well," Grant shrugs, "something else that I'll just have to handle as we go along."

Meanwhile, there are other duties he must perform.

His mind already is occupied with far more important things: finding provisions for nearly 100,000 men and perhaps 80,000 animals; prevent-ing any flare-ups between the two armies; picking up his responsibilities as commander of all the Federal armies.

He's thinking of those things and is well on his way toward his camp when an aide calls him back to the present.

"General, hadn't you better notify Washington of the surrender?"

Grant nods, smiles a little sheepishly, and dismounts. Sitting on a large stone beside the stage road, he writes a message for a scout to carry to the nearest telegraph station.

Headquarters, Appomattox C. H. Va.
April 9th, 1865, 4:30 P. M.
Hon. E. M. Stanton, Secretary of War,
Washington
General Lee surrendered the Army of Northern Virginia this after-noon on terms proposed by myself. The accompanying additional correspondence will show the conditions fully.
U. S. Grant,
Lieut.-General

He's not gone much farther when cannon fire shatters the silence behind him. An alarming sound. Then Grant recognizes its rhythmic ca-dence: a salute. They're firing a salute, celebrating the victory.

"General Rawlins," half angry, he turns in his saddle, "this war is over. Those men are our prisoners, but they're also our countrymen again. We'll not crow over them. Let's have no artillery salutes; no offensive remarks; nothing that will humiliate the Confederate soldiers. See to it immediately."

When he's reached his headquarters' camp, farther along the stage road, he doesn't outwardly show it, but he too is too wound up to do more than simply sit by the fire and consider what has happened.

A light rain has begun to fall, but Grant and his staff thoughtfully sit around their campfire, ignoring it. Each man is wrapped in his own thoughts, but all of them are eager to hear Grant's personal observations about Lee's surrender. Something historic to recall.

Grant, however, just sits there, thoughtfully puffing his cigar, it seems to them an interminable time; silent. Then, taking the cigar from his mouth and looking at his chief quartermaster, General Rufus Ingalls, he asks.

"Ingalls, do you remember that old white mule that...," he names an officer they'd both known during the Mexican campaign, "used to ride when we were in Mexico?"

"Well, yes; perfectly," Ingalls, so anxious to encourage any reflections by the taciturn Grant that he'd have been willing to swear to the exact number of hairs in the mule's tail if that would move things along, answers.

Grant, his thoughts returned to the Captain Robert E. Lee he'd met in Mexico and to the other memories their conversation has triggered, picks up the story Lee had interrupted early in their meeting. The mule and its antics during an excursion the young officers had taken to Lake Popocatepetl are more on his mind than what has happened at Appomattox Court House; and, to his staff's frustration, they'll have to settle for that.

It's not until after supper that he even mentions the surrender, telling them that he hopes that Lee's action will persuade Confederate Generals Joseph Johnston and Kirby Smith and other Confederate leaders to lay down their arms too.

Then he surprises them by announcing that in the morning he'll move Humphreys' and Wright's corps, most of Ord's Army of the James, and most of Sheridan's cavalry back to Burkeville. Ord will send a detachment to pick up the Confederates who'd fled to Lynchburg. Only Griffin's V Corps will remain at Appomattox Court House to complete the surrender. Generals Gibbon, Griffin, and Merritt will oversee the surrender. Rations will be short. Men must double up, sharing what they have with each other and with their surrendered enemy.

As for himself, he doesn't plan to spend any more time than necessary at Appomattox.

"No," his mind is made up, "the job's done here, and I'm needed back at City Point. Then I'll go on to Washington. This war's costing millions of dollars every day; we need to wind it down."

He plans to leave the next day himelf but confides that he'd like to meet with Lee again. He adds, however, that because he's determined to do nothing that could insult Lee or his soldiers, he'll not enter the Confederate camp.

When his headquarters' campfire has burned low, Grant and Adam Badeau sit alone before the fire and Grant finally reflects on the terms he's given Lee.

"Do you know, Badeau, until I sat down at that table with the pad before me, I'd not given a single thought to how I should express the terms of surrender. I was very mindful of the President's feelings on the matter, but I'd not yet thought out how I might state the terms."

"I wish," he shares, "that the President could have remained at City Point so that I could have first reported it to him."

"You know, Badeau," Grant continues to reflect, "some of our officers dislike the idea of paroles. Sheridan, well you know Sheridan," he leaves the thought unspoken, "Well, some of them think that the ranking Confederates, certainly Lee, should have been held for trial."

A pause then, "I just couldn't see that."

Badeau nods, "I agree, General, and I think that the country will too. Your terms will hasten the reconciliation. That's what matters now."

Grant is silent for a long time, puffing his cigar, thinking, then he quietly closes the thought, "Well, I'll keep the terms, no matter who's opposed. And Lincoln's sure to be on my side."

LEE

Lee has ridden east on the stage road, past the courthouse and down the long ridge below the village to the apple orchard where he'd rested that morning.

Between the village and the orchard he passes through some of Gordon's infantry, and word of the surrender spreads fast.

Already men are gathering, talking in small groups; their despair evident. Color bearers, anticipating the surrender, have begun to tear their battle flags and to parcel out fragments to the men who'd fought under them.

Confederate soldiers bend rifle barrels between tree limbs or smash their weapons' hammers with rocks so, "No Yankee'll ever use you against any Southern soldier."

Officers drive their swords into the ground then snap off their blades. Everywhere knapsacks, sword belts, and canteens have been discarded.

As he descends the stage road toward the orchard, the Confederate soldiers nearest the road are the first to see him. Lee, always before erect, confident, assured; now with his head low on his chest, looking straight ahead, over Traveller's ears.

The word quickly spreads: "Lee! It's Uncle Robert!" and more and more of them run to the road to cheer him. Some of his soldiers, still defiant, call, "General, we'll whip 'em yet!" or "General, say the word and we'll go in and fight 'em again."

Most of them, however, knowing what has happened, simply reach out to touch his boots or to pat Traveller as the big gray horse passes. Close to him now, they see the tear-streaked face Lee no longer can control.

Perhaps it's the emotions that he no longer can mask that cause him to stop short of his headquarters' camp and turn again into the little orchard. There he dismounts under the tree that had sheltered him before his meeting with Grant.

He'll be there for more than an hour, protected by the cordon Colonel Talcott again establishes, pacing back and forth, as one staff officer puts it, "In one of his savage moods."

They try not to disturb him unless a courier brings a report or a visitor comes to pay his respects.

Already the visitors are coming. Northern officers determined to see him, to pay their respects to an old friend or simply to be able to say that they *did* see him this day.

He's formally polite but doesn't offer to shake hands with any of them; does nothing to prolong an interview. They're all very deferential toward him, approaching with bared heads and respectful bows, keeping their conversation mercifully short.

Couriers also come and go: reports from his commanders and staff officers. Some wagon loads of food already have arrived from the Federal commissary. Thank God for that.

Near sunset Lee leaves the orchard to ride to his headquarters' camp.

South Carolinian Major E. M. Boykin is among the first to see the lone gray horse and its rider emerge from the orchard, his call striking the Confederate camp like a lightning bolt. Gray soldiers stop whatever they're doing to run to the road; cheering, calling, shouting as they come. Across the valley and up the long ridge beyond, from the very edges of their camp, they come.

Lee, his head bent forward, his gray beard touching his breast, and his hands resting on the pommel of Traveller's saddle, is allowing the slack-reined horse to slowly make its own way down the road.

When the men begin to cheer, however, Lee immediately rouses, looks about, begins to smile. Then he lifts his gauntleted right hand to his hat and, hesitantly at first then surer, waves the hat to the soldiers who now crowd about him. His tanned face, no longer pale but flushed, his eyes blazing: the look they've cheered and followed on so many battlefields there once more.

Traveller moves slowly, sometimes forced by the crowds of men to stop altogether for a moment or so, then continuing. The horse once more enjoying a parade at which he and his master are the center of attention, tossing his head and tail, adding his neighs to the cheers around him. Feeling hundreds of hands gently pat his neck and shoulders and flanks as he passes. Not understanding but enjoying the moment.

The narrow, sandy stage road is packed with Confederate soldiers. Every man bareheaded now, heads and hearts bowed down. As Lee nears they cheer until their throats ache, and their chests grow too tight to breathe.

Then, when he's close enough for them to see his face, they grow silent, thoughtful. So the impromptu parade will be for its entire length: loud cheers, then choking sobs, then thoughtful silence. When he's passed, the same emotional sequence picked up by the next group waiting beside the road.

Those who can speak say "good-bye" in whatever way seems best to them. Those who can't find the right words or just can't say them, simply look into his eyes and pass their hands gently across Traveller's sides.

One soldier reaches out to grasp Lee's hand then says it for all of them, "I love you just as well as ever, General Lee!"

Lee, bareheaded, is erect in his saddle now; the way he knows they expect him to look. The pain in his chest, so much worse that he'd had to rest in the orchard, has returned. But this time it's a different pain: a feeling of something so big his chest just can't contain it. Even breathing is hard. Tears still stream down his face, but the man is erect, confident, unbeaten.

Finally he turns into the little grove of oaks and pines above the road. Marshall is there, anxiously awaiting him. For a moment the colonel fears that he'll have to help Lee from his saddle, but Lee smiles his thanks and nods to reassure him. Marshall returns the smile, realizing that the Old Man still commands himself, and them, and will to the end.

They've found a tent for him, one with Union Army markings. He pauses before it, feeling the need to say something to those nearby. Most of them can't hear him, but those who can later tell the others that he'd said something like, "Boys, I've done the best I can. Go home now and, if you make as good citizens as you have soldiers, you will do well. And I shall always be proud of you. Good-bye, and God bless you all."

It's not enough, and he knows that. Nothing he can do will be enough. But he must try.

"Colonel Marshall, I must rest. But I want you to compose a farewell order for our soldiers. Let me have it as soon as you've finished."

Then he disappears into his tent.

THE ARMIES

Lee is concerned about the armies mingling too quickly lest tempers flare, and Grant has posted guards to keep Billy Yank from crossing into Johnny Reb's lines. Yankee guards don't enforce the restriction, however, and, besides, no one said anything about keeping Johnny Reb at home.

Thousands of the veterans, Blue and Gray alike, simply take the end of the fighting as an invitation to go meet their former enemies, now countrymen again. Most of the resulting meetings are harmless and friendly.

A pair of Confederate privates are sent as couriers to the Federal headquarters. As they rein their horses, Yankee orderlies, not able to distinguish a ragged Confederate officer from a ragged Confederate private, run to take their horses. The Southern soldiers dismount with all the dignity and style they can muster, enjoying their brief moment in the sun, content to know that their army may have surrendered but that they at least have managed to fool the Yankees one more time.

Union soldiers stroll into Major Boykin's 7th South Carolina Cavalry camp, suggesting, "Say, boys, we've been fightin' one another for four years now, and we have a heap of respect for you Johnnies. How 'bout swappin' us some of your five-dollar bills to remember you by?" And they do.

A Confederate entrepreneur inquires, "Hey, Yank, you got any fish hooks?"

"No. Haven't seen one for more'n a year. Any place to go fishin' here?"

"Don't know. Thinkin' 'bout when I get home."

Food's first on their minds. Longstreet personally goes to Grant's Commissary General to ask, "For God's sake, can you send anything at all?"
He does.

An officer of the 33rd North Carolina Infantry Regiment also tries a direct approach at a newly established Federal commissary's supply tent.

"Can you give me some bread for my men? They've had nothing at all for three days."

"I can't do it yet," the reply. "Leastwise not official like. But see here, Johnny; you just walk around the tent casual like and fill your haversacks with crackers and loaf sugar. Yes, and your canteens with some of that whiskey over there. I don't reckon anyone'll bother you 'bout it."

A 1st Massachusetts Cavalry Regiment trooper shares his rations with a grateful Confederate who tries to talk between bites.

"Doggone, that's splendid coffee! First I've had in a long dry spell! You all overpowered us; we just couldn't hold out a day longer on wind alone. Say, I like this meat! I tell you it's good! I didn't know I was so hungry. First time I've had hardtack that weren't wormy in a long time! Must've gone beyond the hungry point."

For hungry Confederates the aroma of real coffee is staggering; butter, condensed milk, sugar are beyond their dreams.

Custer has many guests, Blue and Gray, at his camp; most of them friends from West Point days. They pile rations in an impromptu picnic as Custer's bands play "Hail Columbia," "Yankee Doodle," "Dixie," "Bonnie Blue Flag," and the one that tears at all their hearts, "Home Sweet Home."

Union General John Gibbon is talking with Confederate Generals Longstreet and Heth when another old friend, Confederate General Cadmus M. Wilcox, rides up on a sorry looking old horse. It's a warm day, but Wilcox is wearing a long gray overcoat, buttoned at the throat. Asked if he's cold, Wilcox smiles, shakes his head, and confesses, "No, but this is all I have. Sheridan got our baggage."

Beneath the coat, his only garment is a tattered shirt. Gibbon empties his field trunk for clothes to share with Wilcox.

The encounters, however, aren't all entirely friendly. A Federal colonel feels compelled to deliver a speech to a group of Confederates. In effect, he tells them, "We'll not bother you; we forgive you."

One of his ragged audience listens for a moment then quietly answers, "If I had you out in the woods by yourself, by thunder, I'd bother you."

West of Appomattox, Confederate cavalry General Fitz Lee, temporarily free of Sheridan's cordon, encounters a lone Confederate veteran, unaware of the surrender, hurrying from leave to rejoin Lee's army.

"Never mind, old man," Fitz Lee tells him. "You're too late now. Lee has surrendered. You'd better go home."

"What's that? Lee's surrendered?"

"Yes, that's right."

The soldier stares at him with tear-filled eyes then blurts, "No, sir! You can't make me believe that Uncle Robert has surrendered. It must've been that goddamn *Fitz Lee*!"

A North Carolina officer, not sure how Confederate officers will be treated after the surrender, hands his Negro slave, George, a five-dollar gold piece he'd saved for months. Then he tells him, "Now, George, you take my horse and go home, a free man."

The boy, confused and frightened, thinks about it then blurts out, "Mars Tom, I can't do it. I'se gwine to stay with you. If you can stand a Yankee prison, I can too. When we left home for you to jine the army my Mammy tole me not to ever show my face agin without you was there too. Dead or alive I was to stay with you. I've got to mind Mammy, Mr. Tom. Got to stay with you."

CHAMBERLAIN

For General Joshua Chamberlain, thoughtful, contemplative, the day has been full beyond all measure. He'd seen Lee then Grant enter the brick home just off the stage road. Then he'd watched them leave. But he'd not joined the gathering of Union and Confederate officers before the courthouse.

Most of them are West Pointers, friends from the Old Army, and he knows few of them, none intimately. And despite a combat record they'd all envy, he's always felt a bit apart because he's not West Point, not Old Army.

Besides, he wants time with his own thoughts as he watches the two armies, hours before deadly enemies but already beginning to mingle.

"Ask the men to share what they have, Ellis," he'd ordered. "Set our lines along the crest of the ridge here. Their skirmish lines are a couple hundred yards down the slope, but I expect they'll pull back to the far slope for the night. Put out security, just to keep our boys alert, but I expect that this will be the last night we'll need it. The last night in nearly four years, Colonel! Wonderful, isn't it? I'll go find General Griffin for orders."

———

At twilight he rides to where the stage road begins its descent toward the Appomattox. Just off the road there's a little knoll where, looking back, he can see his own camp and, looking across the valley before him, he can enjoy the sight of hundreds of Lee's campfires blazing in the evening sky.

He knows that men are clustered around each fire: smoking their pipes, clasping their knees, talking about the day and about so many days passed and now the hope of so many to come.

As the night darkens, he watches as sparkling trails of light flash across the sky; riflemen firing impromptu fireworks. Now and then a signal rocket, but more often they're infantrymen using their muskets to fire shell fuzes. Sparkling paths, once deadly but now harmless. A good thing.

The campfires have begun to burn low long before he hears the first bugle sounding "Taps." The soldiers, Blue and Gray, have reason to be tired and, with a little food in their bellies, tobacco in their pipes, and knowing that they can get their first full night's sleep without fear of attack in a long, long time, they're turning in early.

Then the clear, haunting notes of "Taps" echo around him. A single bugler, then another and another repeat the call. One by one, regiment after regiment, camp after camp. Soft, plaintive notes echoing over the fields, echoing from the hills beyond.

"Dear Lord," Joshua Chamberlain looks to the sky and speaks his thoughts, "how beautiful that sound. And how much more beautiful above the campfires of two armies sleeping, not fighting. Not going to fight any more. How beautiful."

Then he confesses, "I love Fannie and the children so much. But I shall miss this, Lord. I shall miss this."

"Taps," the last notes of the last bugler's call seeming to linger, as if reluctant to go, then gone.

Rain begins to fall and somewhere, up the road to his left, he hears Sheridan's bands playing very softly the plaintive notes of "Home, Sweet Home." A benediction on what has happened this day.

CHAPTER 9

The Mingling   April 10–11, 1865

GRANT

Grant is up early. Through the night the rain drummed hard on his tent, and at dawn it still drizzles. A cold rain.

Nearly 100,000 men are camped nearby. His own men are reasonably sheltered from the weather; Lee's army is exposed to the worst of it.

"Well," he sighs, "at least it was the first night of peace. Thank God for that. I wonder if Lincoln knows. Perhaps; if not, he soon will."

About nine o'clock, wanting to talk with Lee again before he leaves for City Point, he rides to the Confederate picket line just off the stage road below the village. He doesn't plan to enter the Confederate camp, and that's just as well for ragged Gray pickets, still carrying their rifles but probably without a single round of ammunition among them, stop him. Courteous but firm.

"If you'll wait, sir, I'll send word to General Lee."

Grant, amused at Sheridan's reddening face and scowl, nods politely.

"If General Lee's duties will permit, Captain, I would appreciate his joining me."

Lee, in his rain-soaked camp, also has been up and busy for some time. Colonel Marshall hasn't finished drafting Lee's final order for his army, and he reminds him that he's anxious to have it.

Other matters are more pressing. Grant's quartermaster wagons continue to bring provisions into his lines. Not nearly enough, but the most they can spare, and he's grateful for it. Grant's quartermaster, however, has little forage for his own horses, let alone for Lee's. Doubly painful for Lee, the lover of horses. He knows that the Confederate army's horses are starving and that, even if forage is provided, many of them won't recover. Sad.

He's discussing that with his staff when word comes that Grant is at their picket line. Embarrassed that he'd not been there to greet him, Lee at once mounts Traveller and hurries up the stage road.

Grant, sensing Lee's embarrassment, lifts his hat in friendly greeting and, around his perpetual cigar, smiles, "Good morning, General."

Grant's staff also salute, and Lee returns the courtesy, "Good morning, General."

Then the two commanders ride a bit ahead of the others, to the little knoll overlooking the hundreds of Confederate campfires scattered across the valley and the slope beyond it; their staffs dropping a bit behind as the two generals talk privately.

There are preliminaries, Grant assuring himself that provisions have begun to reach Lee's army.

"We'll do better today," he promises. "The Military Railroad has gotten to Burkeville. The roads between there and here, however, are very bad. We're working on them, and I've ordered our engineers to repair the railroad tracks. Once that's done, we'll get rations and forage forward. I'm more concerned about forage at the moment. We have very little on hand for our own horses, let alone for yours."

"Yes," Lee understands. "Your men have been kind to share their rations. I'm afraid that many of our horses are nearly gone. I've sent details out into the countryside in hopes of bringing in some forage but," he reflects, "it's a poor country."

Then Grant turns to other matters.

"General, what we've done here will go a long way toward ending this war. And I know that nothing would please President Lincoln more than to have it lead to a general peace."

"Yes," Lee nods. "For us the issue has been decided; and I too want no more loss of lives. But the South is a large country and it's people very independent. Your army may have to cross it three or four times before the last resistance ends. But it must end, and I pray soon."

"General Lee," Grant quietly suggests, "no one has greater influence among the people of the South than you. If you would suggest that the other armies lay down their arms, they'd do it."

"Perhaps," Lee answers. "It's kind of you to say so. But," he politely declines Grant's suggestion, "you and I are soldiers. We can't advise any other commander to lay down his arms. We can only recommend to presidents and abide by their decisions."

"Of course," Grant doesn't press the issue.

"In any event," Lee completes the thought, "the war cannot go on much longer. That is clear."

They talk quietly, pleasantly about half an hour before Grant turns to leave. He'll not see Lee again at Appomattox.

Grant returns to the McLean house, establishing himself in a chair on its porch where he smokes his cigar, receives couriers, and greets other Union and Confederate leaders as they come and go.

Inside the house Confederate Generals Longstreet, Gordon, and Pendleton and Union Generals Gibbon, Griffin, and Merritt work to carry out the surrender terms Grant and Lee had reached the day before.

When Longstreet comes to the porch, Grant's particularly glad to see his long-time personal friend and best man at Grant's wedding so long ago.

When he sees Longstreet, Grant rises from his chair, slaps Longstreet on the shoulder, and exclaims, "Well, Pete, I wish that we could play a game or two of Brag as we did in the old days."

He finds a cigar for him and, for a time, Longstreet sits beside him on the porch, smoking and sharing reminiscences.

Then Grant mounts Cincinnati for the ride to Burkeville. When he arrives there, his train is not yet ready to leave so he uses the time to visit a nearby Federal field hospital; a duty he always dreads.

This visit is no easier for him than all the others he's made to military hospitals. It's filled with wounded and ill Union and Confederate soldiers. Some of the patients, even though wounded in the last of the fighting, are happy to have survived and to know that they'll soon be on the way home. Others, badly hurt, grope for some reason for it.

A young surgeon takes the opportunity to report that yesterday had been very bad for them.

The hospital was nearly full, with wagons and ambulances still arriving with men wounded in the fighting at Appomattox. Supplies of candles, bandages, and opium were low. Rain poured down; and exhausted surgeons and attendants, covered with blood, worked through the day.

The tired surgeon reminds Grant that even as the North rejoiced in their victory at Appomattox, he'd watched soldiers paying its price. Grant, who needn't be reminded, nods his understanding.

Then the surgeon brings another matter to his attention.

"Late in the afternoon, General, we'd put a thousand wounded men on the hospital train for City Point. It was supposed to leave right away, but it got dark and the train still sat there. When I complained, I was told that they couldn't find the doctor in charge, and couldn't release the train without him."

"It was nearly midnight," he angrily continues, "when we found him, drunk and playing cards in a house a half mile from here."

"It was 2 a.m. before that train got started, General. And nothing done to the doctor. Drunk or not, they just put him on the train and started out. Doesn't seem right, sir; doesn't seem right."

When the young doctor has finished his report, Grant nods his understanding then turns to Rawlins.

"General, send a wire to City Point. See if we can't get a special train out here right away with medical supplies. Then it can take an extra load of wounded back."

He takes the cigar from his mouth and points for emphasis, "And find that doctor. Put him under arrest. If this is true, I want him court-martialed. And, Rawlins, do what you can to see that this doesn't happen again."

Finally his own train gets under way. The rails of the army's "Washboard Railroad," however, are so loosely laid that three times the train jumps its tracks, and it takes twelve hours to make the sixty-one mile trip.

His train limps into an empty City Point station near dawn.

"Sorry, General," the stationmaster reports. "Mrs. Grant was here waiting for a long time. Said she'd planned a special dinner for you; reckon you're in trouble for that. Anyways, when we got word of the third derailment, she gave it up and went back to the *Mary Martin*."

He nods his understanding but is more concerned that their horses be safely unloaded than about the absence of a welcoming party.

"Rawlins, do you suppose they'll be able to unload the horses all right?"

"Never mind, Grant. You go on and find Julia. I'll see to it."

He hurries up the *Mary Martin*'s gangplank to surprise her. Her special dinner, heated again, makes a fine breakfast.

He soon returns to his headquarters, however, and spends the day there, busy with many things.

Once, Porter interrupts his work to suggest that he go see the city he'd besieged for nearly a year.

"No," he shakes his head. "It's better that I not go up to Richmond. I could do no good there, and some of the people would resent my presence. You go on; tomorrow we'll be going back to Washington."

His mind already has turned from Appomattox to the job of winding down the war elsewhere.

LEE

After his morning meeting with Grant, Lee is returning to his headquarters when he encounters a Federal general he doesn't recognize until the stooped, bespectacled officer waves his hat and calls, "Good morning, General."

"General Meade? George, is that you? I almost didn't recognize you. But what in the world are you doing with all that gray in your beard?"

Meade wryly smiles, "I'm afraid, General, that you have to answer for most of it."

Lee and Meade, old friends, ride together to Lee's headquarters.

As they ride, Confederate soldiers, seeing Lee, again begin to cheer. Meade, thinking that they have mistaken him for a Confederate officer, orders his color bearers to unfurl the Federal flags they carry.

A grizzled Gray veteran, seeing the Stars and Stripes suddenly waved in his face, makes his own position clear: "Damn your rag, General; we're cheering for General Lee!"

Lee begins to apologize for his soldier's behavior, but Meade's irritation that he might have been taken for a Confederate has gone.

"Old soldiers, General Lee," he chuckles. "No matter the color of their uniform or their rank, they're pretty much alike in what matters to them."

Later, as they talk, a curious Meade asks him, "Tell me, for I've often wondered, when we finally were able to break your lines at Petersburg, about how many men did you have?"

"About 33,000 on our rolls, but that was our entire army, and many of them were not in the lines facing you. And you?" the question is one professional's to another.

"A little more than 50,000 in my sector alone."

Another friend, Grant's Chief of Artillery, Brigadier General Henry Hunt, also visits Lee.

He tells Lee that he's just chided his former artillery student, Confederate General Armistead Long, for not better coordinating the Confederate artillery barrage before Pickett's disastrous infantry charge at Gettysburg.

"Yes," Lee nods. "You are correct. Our artillery did not do well there; but it was more my fault than General Long's. He did not have the authority to do all that he might have done."

Perhaps it's these professional exchanges that prompt Lee to begin his final report to Jefferson Davis.

He asks the comments of his commanders and staff and, when he studies them, he's certain that he'd made the right decision in surrendering his army.

When he met with Grant his army numbered about 27,000 men, but less than 10,000 of them still carried weapons. And his supply of ammunition for them was very low: about seventy-five bullets per man, and none in reserve.

His sixty-three remaining cannon were down to less than two shells per gun.

His cavalry, mounted on horses so weak from hunger and exhaustion that many of them staggered beneath their riders, numbered less than 3,000.

Had he not ended the fighting, many more men would have been killed or wounded and, at best, it could only have delayed the surrender a few hours.

He begins his report to Davis: "It is with pain that I announce to your Excellency the surrender of the Army of Northern Virginia." It will take him the better part of this day and the next to complete it, but it's a duty he feels necessary.

Late in the morning he interrupts his work to return to another matter very important to him.

"Colonel Marshall, how are you coming with our final order?"

"General," Marshall is embarrassed and frustrated, "I've been interrupted so many times that I've not been able to complete it."

Then he confesses, "No, sir; that's only partially correct. The truth is that I find it very hard to write. What can I say to men who've given so much?"

Lee nods, kindly but firmly. Then he points to his ambulance-office.

"Go into the ambulance, Colonel. I'm going to post a guard there so you won't be disturbed again. Find the words, sir! Find the words! Come to me when your work is done."

Late in the afternoon Marshall brings a draft of the order for Lee's approval. He reads it carefully, thoughtfully, several times.

"This is very good, Colonel, but I think that we should strike these few phrases," and he does so with his pencil. "They might cause bad feeling; we don't want that."

There are several other minor changes then Marshall returns to the ambulance to complete his work.

When Lee has approved the final draft, Marshall hands his pad to a clerk.

"Here, Bell. Copies for each corps commander and for our own records. Do it carefully; this will be seen by a great many people."

The order, the last Lee will authorize for the Army of Northern Virginia, will become famous in military history.

GENERAL ORDER, NO. 9
Headquarters, Army of Northern Virginia
April 10, 1865

After four years of arduous service marked by unsurpassed courage and fortitude, the Army of Northern Virginia has been compelled to yield to overwhelming numbers and resources.

I need not tell the brave survivors of so many hard fought battles,
who have remained steadfast to the last, that I have consented to
this result from no distrust of them; but feeling that valor and devotion could accomplish nothing that could compensate for the loss
that must have attended the continuance of the contest, I determined to avoid the useless sacrifice of those whose past services
have endeared them to their countrymen.

By the terms of the agreement, officers and men can return to their
homes and remain until exchanged. You will take with you the sat-
isfaction that proceeds from the consciousness of duty faithfully
performed; and I earnestly pray that a merciful God will extend to
you His blessing and protection.

With an unceasing admiration of your constancy and devotion to
your Country, and a grateful remembrance of your kind and generous consideration for myself, I bid you all an affectionate farewell.

(Sgd) R. E. Lee
Gen'l.

Commanders read copies of the order to their soldiers, and they take
it to their hearts. Many of them copy it then come to have the copy signed
by Lee, a priceless heritage.

One who comes to Lee's tent is Colonel Herman Perry, who'd re-
fused Union General Seth Williams' flask of brandy when Williams carried
Grant's first surrender message to Lee.

Perry, with a carefully written copy of Lee's General Order Number 9,
has come to ask that Lee sign it. Later, he'll share why the document means
so much to him:

It is the best authority, along with my parole...why after that day I no
longer raised a soldier's hand for the South. There were tears in his
(Lee's) eyes when he signed it for me, and when I turned away
there were tears in my own....He was the greatest man who ever
lived, and as a humble officer of the South, I thank heaven I had
the honor of following him.

Lee's final order, more than any other thing, marks the sunset of the
Confederacy.

In the evening of this full day, the band of the 4th North Carolina Infantry Regiment comes to serenade Lee. When they play he appears before his tent to listen. When they've finished, he smiles and thanks them, "God bless you, men! I can say no more."

THE ARMIES

Throughout the day six officers: Gibbon, Griffin, and Merritt representing Grant; and Longstreet, Gordon, and Pendleton representing Lee, draft specifics of the surrender agreement between the two generals.

They begin their work at the Clover Hill Tavern but soon abandon the bare and cheerless building for the McLean home. In the same parlor where Grant and Lee had met the day before, they work out details of the final peace arrangements.

One of their early problems is to agree upon which Confederate units are included in the surrender. Does it include only the Confederate troops around Appomattox Court House or does it also include the cavalry and artillery that escaped Sheridan's cordon? And what of the stragglers and deserters, still being picked up by Federal patrols, miles from the village?

Longstreet, quietly puffing his pipe, suggests, "Make it twenty miles. Everybody within twenty miles is included in the surrender. Those beyond that are on their own." Good enough.

Following the formal laying down of their arms, their flags, and their equipment, Lee's officers and men will be issued papers attesting that:

> We, the undersigned prisoners of war belonging to the Army of Northern Virginia, having been this day surrendered by General Robert E. Lee, C. S. Army, commanding said army, to Lieutenant General U. S. Grant, commanding Armies of the United States, do hereby give our solemn parole of honor that we will not hereafter serve in the armies of the Confederate States, or in any military capacity whatever, against the United States of America, or render aid to the enemies of the latter, until properly exchanged, in such manner as shall be mutually approved by the respective authorities.
>
> The within named officers will not be disturbed by the United States authorities so long as they observe their parole and the laws in force where they may reside.

Gibbon's headquarters has a small press they put to work printing 30,000 copies of the paroles. Then they must be distributed to the corps, divisions, brigades, and regiments of Lee's army where officers will sign each one then issue them to their men.

The issue of Confederate soldiers' surrendering their arms and their flags comes closest to causing a problem. Longstreet, Gordon, and Pendleton, hoping to avoid the humiliation of having the Confederate soldiers lay their weapons and flags at the feet of their conquerors, ask that their men be allowed to simply leave them in their camps for Union soldiers to collect after the Confederates have gone.

"Laying down our arms and our battle flags is a sensitive issue," they suggest. "Can't we do it without the confrontation that a face-to-face meeting might bring on?"

Gibbon takes it to Grant, waiting on the porch, who quietly shakes his head.

"We don't want to humiliate them, Gibbon. But we're due something more than our simply picking up what they've left behind. No, I want them to march in, lay down their flags, their arms and their equipment; then withdraw. That's all, but at least that. And, it seems to me that Chamberlain would be fine for the ceremony. He'll handle it with the touch that's needed."

So it's decided.

When they bring the final document for Grant's approval he adds a note authorizing free transportation on Northern military railroads and steamers for Confederate veterans returning to their homes.

That's a start, but it doesn't include the civilian railroads and steamers veterans will need to reach the farthest extent of the Confederacy: Texas, Louisiana, Mississippi, Florida.

Gibbon points this out to the Confederate commissioners, suggesting, "Your men will have no money for the transportation they'll need down there."

"Never mind," Longstreet gestures with his pipe. "We can handle that among ourselves. 'Sides, once our boys get a ways down the road, our own people will be the conductors on those trains. When they find out what's happened, they'll know what to do."

The Northern commissioners don't press the issue. Had they done so they might have learned that awhile back Fitz Lee had captured several Yankee paymasters' wagons, and so a lot of Federal greenbacks quietly rest at Lee's headquarters. Gordon, Longstreet, and Pendleton have decided that the greenbacks were taken fair and square in battle, before the surrender took place, and are the legitimate property of the Confederacy. So they see no need to even mention them to Grant's commissioners; they'll just divide them among their soldiers.

When they've finished their work and adjourned, Gibbon sits alone in the stripped-bare parlor of Wilmer McLean's home. He too would like a souvenir of the occasion but little remains. For their work they'd improvised a table, his own campaign-scarred, wooden camp table, draped with an old army blanket. Shrugging his shoulders, Gibbon asks his clerk to sand the table's top and have it inscribed.

The commissioners have decided to space out the surrender ceremonies for the various elements of Lee's army. As the horses of the cavalry and artillery suffer most, those units will lay down their arms and flags first, then be paroled.

On the eleventh of April the remaining Confederate cavalrymen, nearly 1,600 of them, surrender to Federal Major General Ranald Mackenzie, take up their paroles, and leave. Then 2,576 Confederate artillerymen also lay down their flags, their guns, and their equipment. They must leave most of Lee's cannon on the stage road below the village because their exhausted horses can't pull the guns up the long hill. Union quartermasters there will find Confederate horses dead within their traces or beside the road.

Grant's ordnance officers will return 151 captured Confederate cannon to City Point: 61 surrendered at Appomattox Court House; 23 more taken by Custer at Appomattox Station; 13 concealed in nearby woods; and 54 more found buried at Red Oak Church, eight miles from the Court House. One of Major Young's scouts had helped bury them there and, following the surrender, helps dig them up again.

One act of the drama remains. Lee's infantry, once the finest in the world but now totally decimated, must march before Joshua Chamberlain, now promoted to major general, and his ranks of Union soldiers to lay down their arms and their battle flags. The ceremony will take place the morning of the twelfth.

Chamberlain leaves the smaller 1st Brigade to resume command of his old 3rd Brigade, 1st Division, V Corps; while his senior, Major General Joseph Bartlett, commands the division. Because Grant has designated Chamberlain to receive the surrender of Lee's infantry, Chamberlain will command the entire division during the surrender ceremony, General Bartlett graciously stepping aside for it.

Chamberlain, recognizing the slight to Bartlett; that Meade, Ord, Gibbon, and most of the other commanders and soldiers who'd done so much to make the day possible won't be present; and hoping to spare Lee's army as much humiliation as possible, determines to make the ceremony a simple one.

As their officers make plans to complete the surrender, Union and Confederate soldiers, still on short rations, are slowly recovering from the hunger, cold, and exhaustion of the week before.

Most of the Confederate infantrymen, now resigned to the inevitable, spend the two days they must wait for the surrender ceremony and their paroles sitting about their campfires or trading souvenirs in Yankee camps.

Even the apple tree where Lee rested before and after his meeting with Grant isn't spared. Confederate soldiers cut it down, break its branches into little pieces, and even sell individual leaves as souvenirs; $5 per chip the going price.

When the tree, its branches, and its leaves are gone, they dig out its roots and sell those too. Finally, only a hole remains where the tree had been the day before.

Other enterprising Southern infantrymen cut down several other nearby trees and peddle their branches in the Yankee camps, solemnly attesting to their coming from the very tree under which Lee had sat.

During their last evening in the field, John Gordon walks among his regiments, talking with his officers and men.

Soldiers are gathered around hundreds of campfires singing, praying, joking, retelling their experiences, and trying to imagine being able to start for home the next day.

As they talk among themselves only a few show ill will toward their conquerors. And there is no suggestion of guerrilla warfare. Instead they see the war they've fought for four years as a question that's been fought to the finish. They've done the best they can, and now it's time to move along; to rebuild their lives.

Meanwhile, their officers, their work lit by scraps of tallow candles fixed to the sockets of bayonets jammed into the ground, continue to sign the paroles that will make their leaving possible.

CHAMBERLAIN

Like John Gordon, Joshua Chamberlain spends the evening at the campfires of his soldiers.

Conflicting emotions tug at his heart: pride in his soldiers but sympathy for their fallen enemy, now countrymen again; gratitude that the fighting is over but with the passing of the armies a sense that the high point of his own life has passed too; longing for his family but doubt that he can be content with a classroom again.

Most of all, however, he's keenly aware that the men with whom he's shared so many long marches and hard-fought battles soon must go their separate ways, probably never to meet again. Happiness, but already a loneliness.

So this night, their last night camped in the field before the Confederate army, he's drawn to his V Corps' campfires.

In the late afternoon he visits his old command, the 1st Brigade, 1st Division, soldiers he'd led during the hard fighting before Five Forks then the pursuit to Appomattox Court House.

Officers and sergeants of the 185th New York Infantry Regiment quickly recognize the tall, slender Chamberlain as he nears their campfires. They cheer, appreciating his coming, and welcome him.

"Has the General had supper, sir?"

"No, not yet, Sergeant Major."

An orderly soon returns with a tin plate heaped with food and a tin cup of steaming coffee.

They talk about the surrender ceremony and the concern uppermost in their minds, "How long, do you reckon, General, before we'll be mustered out?"

Everyone agrees that, with the surrender of Lee's army, the remaining Confederate armies will quickly lay down their arms too. Then the war will be over and they can go home.

Their conversation turns, as it sometimes will with veterans, to a hard fight they've shared. For the 185th Pennsylvania it's the fight for the sawdust pile at the Lewis Farm on Quaker Road, just before their final victory at Five Forks. The fight and the strange sight of their general, Joshua Chamberlain, hatless, covered with dirt and blood, his uniform torn, on a strange, muddy, pale-white horse they'd never seen, waving his sword and rallying the men.

A desperate situation then, but now something they can talk about, even laugh about, and tuck away in their hearts to take out now and then and share with their children and grandchildren.

They recall passing in review before their near-naked general as he dried himself before a campfire after he and Charlemagne took their impromptu swim on the road to Appomattox.

Already their bad memories have begun to give way to the good ones they really want to remember.

Chamberlain moves on to the 198th Pennsylvania, sharing their thoughts about how quickly their prospects have changed.

For most of them the sharp edges of their worst recollections already have begun to soften. For Joshua Chamberlain, some images still are too vivid, too painful. For him the healing process, learning to live with those memories, will take awhile.

He thinks of General Horatio Sickel, lying with a mangled arm among his soldiers, knowing that the surgeons are waiting to amputate that arm but insisting that his men be treated first.

Sickel, desite his own pain, understanding the despair Chamberlain felt that night for all the soldiers, North and South, who'd fallen that day; trying to comfort him.

Chamberlain's mind returns, as it will again and again for the rest of his life, to the image of his subordinate and friend, Major Edwin A. Glenn, fording the chest-deep stream by the Quaker Road against heavy Confederate fire, but taking time to turn and grin at Chamberlain before urging the 198th Pennsylvania forward.

Then, so soon afterward, the image of a torn field above Five Forks and four soldiers carrying a bloody blanket to lay at Chamberlain's feet. When he'd knelt down, a dying Major Edwin A. Glenn's last words, "I did as you commanded, sir."

In the early dusk he turns to the regiments of the 3rd Brigade, the brigade he'd commanded before his near-fatal wound at Petersburg the summer before and now returned to him.

Maine, Massachusetts, Michigan, Pennsylvania soldiers, each different yet alike in so many ways. He feels at home with them.

As the night deepens, he reaches the regiment closest to his heart: his old 20th Maine. The 20th Maine, where he'd first worn his uniform and commanded men, is family to Joshua Chamberlain.

Colonel Ellis Spear, Chamberlain's Executive Officer at Gettysburg when Chamberlain commanded, "Fix Bayonets!", is sitting by their fire.

So too is Chamberlain's brother, Captain Tom Chamberlain. Tom Chamberlain, once a sergeant in the 20th Maine then a young lieutenant, fighting beside him that same desperate Pennsylvania July afternoon. Tom Chamberlain: Joshua Chamberlain recalls having to scold his younger brother that day for calling him "Lawrence" before the men and for standing too near him during the worst of the fight lest their mother lose two sons.

And Sergeant Andrew Tozier, Colonel Joshua Chamberlain's Color Sergeant that day is there. Tozier: badly wounded but standing braced against a large granite boulder at the end of the regiment's line, waiting another charge by the 15th Alabama. Tozier: cradling their regimental flag in his left arm as he raised his Enfield rifle for one more shot at the charging enemy below them.

He's grateful that the memories the Maine men share this evening are light, pleasant ones; mainly about when the regiment and they were all very young.

"General, do you mind Camp Mason, back there in Portland, nigh on to three years ago now, and your first look at the 20th Maine? What did you think?"

"Well, I don't recollect for sure, and of course I didn't know much more about what was going on then than you did; but I expect I may have been like the farmer who was standing in a field with a length of rope in his hands, wondering whether he'd found a rope or lost a cow."

"The one we should ask," he continues the thought, "is old General Adelbert Ames. Thank heavens we had him," and Chamberlain smiles at the memories which come to his mind.

"Adelbert Ames," he continues. "There he was, fourteen months out of West Point, wounded at Bull Run and already brevetted a couple times for gallantry in action. Regular Army to the core, and you all know what that means. Detailed to command the 20th Regiment of Infantry, Maine Volunteers. A brand new regiment; and his Lieutenant Colonel Executive Officer, a college professor who didn't know one end of a musket from the other."

"Later on, when he'd teach me in the evenings and the old 20th had softened his Regular Army ways a bit, he shared some of it with me. His first day, when he walked around the camp, the men paid him no mind at all or, if they did see him, instead of saluting they'd look up from where they were sitting and, real friendly like, say, 'How d'ye do, Colonel?'"

When they've laughed, he continues.

"We didn't have our uniforms yet and we were right there in friendly, old Portland, but of course he had us mount a guard. Regulations said so. Well, you remember the boys took that to be a special occasion, and when he inspected us he found his Officer of the Day wearing a brown cutaway coat, striped trousers, and a silk hat; waving a ramrod he'd found somewhere for a sword."

"I expect that we thought that we looked pretty fair, but he didn't agree. Halfway through that guard mount he'd had enough of our slouching and yelled at one soldier, 'For God's sake, man; draw up your bowels!'"

"Do you remember our first parade, General?" Spears chuckles.

"I remember that band!" Chamberlain grins. "We'd just drawn our uniforms. We must have looked awful, but we thought we looked pretty good. Couldn't keep step at all. Left, right was a little too much for some of the boys so Ames had them tie hay or straw to their feet and the sergeants used 'hay foot, straw foot' until they got the idea. Anyway, Adelbert Ames had the Sergeant Major form a band for that first parade. The men could play their instruments, well, more or less play them, but they'd never practiced playing *together*. Each man was pretty much on his own. Being Maine men, I expect they preferred that anyway."

"The Adjutant got us into some sort of line of companies, and Colonel Ames was about to have us pass in review, when the band took it on their own to start things off. Came sauntering down the line, each man tooting away to his own time, never mind what his neighbor was doing."

"Ames," they all laugh at the thought, "yelled for the Adjutant to, 'Have that man stop beating that confounded drum; have all of them stop that noise.' The trouble was, with all that noise, they couldn't hear the Adjutant so they just kept tooting and beating away."

"Then Ames got mad," Chamberlain's shoulders shake at the thought, "drew his sword and charged the band, scattering them. I expect," he concludes with a grin, "that that was the shortest parade this old regiment ever knew."

"Adelbert Ames," Chamberlain thoughtfully recalls, "definitely was not a mild-mannered man. Once, when we'd done something wrong for the fourth or fifth time, he turned to me and said, 'Chamberlain, this is a *hell* of a regiment!'"

"But you know," he concludes, "after the Gettysburg fight, he came to me and said, 'Chamberlain, this is a hell of a *regiment!*' And he was right, gentlemen, he was right!"

He stands and they all come for a last handshake before he and Tom Chamberlain return to his headquarters.

Tom Chamberlain doesn't call him "Lawrence" before the others now, and Joshua Chamberlain's last thoughts before he sleeps are of how far they've all come, and that tomorrow will mark the end of this part of their lives.

At last "Taps" again sounds across the valley; repeated again and again, from regiment to regiment, until its notes finally fade away. Their last night in bivouac. The last night they'll hear that call under arms. Beautiful. Yet, when it's done, it's hard for them to rest as the bugle call commands. Their minds crowded with many thoughts, most of them don't sleep well.

DAVIS

Miles to the south, Jefferson Davis, having heard nothing from Lee or about his army since Lieutenant John Wise's report, has decided to leave Danville for Greensboro, North Carolina, and Johnston's army. Before leaving, however, he sends a not very realistic telegram which he hopes somehow will reach Lee:

...I had hoped to have seen you at an earlier period....The Secretary of War, Quartermaster General, Commissary General, and Chief

Engineer have not arrived...We have here provisions and clothing for your army....

You will realize the reluctance I feel to leave the soil of Virginia and appreciate my anxiety to win success North of the Roanoke. The few stragglers who come from your army are stopped here...generally without arms....I hope soon to hear from you at this point....

Uncertain that the telegram will reach Lee, he asks young Lieutenant John Wise to return to Lee with a copy. Wise sets out but, along the way, learns of Lee's surrender at Appomattox and turns back to join Johnston's army in North Carolina.

LINCOLN

While Grant and Lee were riding to meet at Appomattox Court House, the *River Queen*, bearing the Lincoln party, entered the Potomac. It's been a rough voyage, the steamer's deck tossing and swept with spray.

Lincoln, entertaining his guests in the ship's saloon, reads selections from Shakespeare's *Macbeth*. He returns several times to Macbeth's soliloquy after the murder of Duncan, several times pausing to reflect on the picture Shakespeare presents of a murderer's mind when, the murder done, he already envies his victim's calm sleep. Preoccupied with the thought, Lincoln re-reads the speech:

> Duncan is in his grave;
> After life's fitful fever he sleeps well;
> Treason has done his worst; nor steel, nor poison,
> Malice domestic, foreign, levy, nothing
> Can touch him further.

Still preoccupied with the thought of death, he recites poetry, solemnly intoning a few lines of Longfellow's *Resignation*:

> ...The air is full of farewells to the dying,
> And mournings for the dead...
> We see but dimly through the mists and vapors;
> Amid these earthly damps
> What seems to us but sad funeral tapers
> May be heaven's distant lamps.

Then, realizing the affect of his mood upon the others, he excuses himself, walking from the saloon to stand alone at the *River Queen*'s rail.

As they pass Washington's home at Mount Vernon, the Count de Chambrun, comes to try to cheer him.

"Mount Vernon, with all its memories of Washington, Mr. President, will be no more honored in America than your own home in Springfield."

Lincoln, mindful of the Count's frequent flattering of Mary Lincoln, turns from his gazing at Mount Vernon to nod, "Yes. Springfield! How glad I'll be to get back there, back to peace and tranquility."

Finally they reach Washington's dock. As they board their carriage, Mary Lincoln looks ahead to the city, shivers as if suddenly cold, and angrily shares with him.

"Washington, Father. This city is full of enemies."

Lincoln, roused from his reverie, turns on her, "Enemies! Mother, we must never again use that word!"

A little past nine o'clock in the evening, Lincoln's Secretary of War, Edwin Stanton, hurries to the White House with Grant's dispatch announcing Lee's surrender. Lincoln and Stanton, so often at odds on the conduct of the war, hug each other in pure joy.

Meanwhile, William Crook, Lincoln's bodyguard, is walking with young Tad Lincoln when the streets suddenly come alive with people, all very much excited. All over the city church bells begin to ring, and they see men and boys waving torches, shouting, beginning impromptu parades.

Crook hails one of the celebrants.

"What in the world has happened? Why all this excitement?"

The man, not recognizing the big former policeman or the President's son, looks at him in amazement. Then he laughs, "Why where in the world have you been? Haven't you heard? Lee has surrendered to General Grant at some place called Appomattox."

CHAPTER 10

Honor Answers Honor   April 12, 1865

CHAMBERLAIN

Joshua Chamberlain lies wide awake in the early morning darkness, listening for rain on the canvas. None. Good. That's an answer to a prayer. And the tent's sides no longer strain at their ropes. Wind's died, that too a good sign for the day.

It rained hard after midnight. He'd been awake then too. Too excited to sleep. And the wound a near constant ache. He'd had some rest, but it would take longer to undo the miles he'd walked.

He smiles now, recalling Sergeant McDermott's last shot, three days before, in the campaign the sergeant has waged, "Sir, would the General mind steppin' under this tree for a word?"

When general and sergeant were out of earshot of the staff, "What is it, Mac?"

"Sir, we've come a far piece. Why won't the General mind just this once and ride the damned horse?"

"The men don't have horses, Mac."

"No, sir, but they ain't generals, and most of 'em are carryin' their Minie balls in their bullet pouches, 'stead of in their hips. Ride the horse, General. Lord knows you've earned the right."

Chamberlain had grinned, "All right. Pretty soon. Another mile."

McDermott had sighed, gone muttering off to root out the brigade bugler. Soon flankers and skirmishers went out, sergeants prodded tired men back to the road, and the 1st Brigade again took up the chase.

Now, turning on his camp cot to ease the ache in his side, Chamberlain concedes, "You were right, Mac. The General should've ridden the damned horse."

From long habit and without looking at his watch, he knows that it's time. He lies quietly, however, playing out the ritual the general and the sergeant established after Gettysburg nearly two years before.

The swaying light of an approaching lantern brightens the tent flap. The canvas parts and a large, dark figure squeezes through the narrow, wet opening, all the while softly muttering about the rain, the cold, the Johnnies still to be surrendered, the Virginia mud, the sentry who'd dared challenge him and near caused the general's coffee to get cold.

Chamberlain lies still until the dark figure approaches, shining the lantern in his eyes and shaking his shoulder.

"I'm awake, Mac. Heard you coming a long way off."

"Mornin', General, mornin'. Time, sir. Four in the mornin', and the rain's stopped. Coffee, sir. Hot as I could make it what with that thick-headed mick's haltin' me out there. Not a Maine man, sir. Probably some Dutchman from the 118th."

As McDermott bends over, he spills cold water from his slouch hat onto Chamberlain's face, and scalding hot coffee sloshes from the tin cup onto the general's outstretched hand.

"Mac, Mac. Did I live through this war to have you kill me now? The day Lee's surrendering his infantry to the old 20th Maine?"

"Beggin' the General's pardon, sir. It's the day we've been waitin' for all this time. Praise the Lord."

"Amen to that, Mac. God's been good to us. But it'll be a hard day for the Johnnies. They won't like laying down their muskets and battle flags. Mustn't forget that."

"Sir, Colonel Spear told me that General Sikes said the General's got the heart of a lion, the soul of a woman. Reckon he hit it right. Now," he points to the still sleeping figure in the other cot, "shall I root out the lad? He'll have to be gettin' back to the 20th. The Sergeant Major won't be waitin' for him or anyone else this day; not this day."

Chamberlain grins, reaches down to grasp a newly polished boot, and throws it at the blanket pulled over his younger brother, Tom.

"Up, Captain Chamberlain; up you come and back to the old regiment. We must be there by sunup. It's a great day for Maine, and we don't want to be late for that."

Several hours later the morning sun, just cresting the hills to the east, has begun warming the air as Chamberlain rides his division's line, a final check on arrangements for the ceremonial surrender of the Confederate infantry.

Chamberlain's lines of Union soldiers begin at the bluff above the river then run up the stage road, past the courthouse, to stop near the McLean house where three days ago Lee and Grant ended the fighting.

About 6,000 officers and men, veterans of thirty battles; nine regiments culled from an original fifteen. Regiments mustering every man who can stand, some wearing bandages but refusing to be excused from the ceremony. Not from this ceremony.

He's placed his 3rd Brigade on the south side of the narrow road, the smaller 1st Brigade behind them, and the 2nd Brigade facing them from across the road.

He rides between the long lines of Northern soldiers, checking, correcting, stopping often to exchange a handshake and a brief word with commanders and sergeants major. Men he knows well, as only an infantry soldier can know those who've fought beside him.

Beginning at the right is the 32nd Massachusetts Infantry Regiment.

"Morning, Colonel Cunningham."

"Morning, sir."

Then the Maine Sharpshooters and, to their left, his old regiment, the 20th Maine Volunteers. He pauses a bit longer at the flags of the 20th Maine, shaking hands with their commander, Lieutenant Colonel Walter Morill, a young captain at Gettysburg; with his brother, Captain Tom Chamberlain; and with Color Sergeant Andrew Tozier. A proud day for all of them.

Next are his Michigan regiments, the 1st and the 16th. Steady, quiet, reserved Midwesterners. Good to have around in a fight. Small regiments; like the others they've seen heavy losses and few replacements. Scanning the hardened faces, bronzed by days in the sun, wind, and rain, Chamberlain knows that the regiments are what remain of brigades at the start of the war. Torn units consolidated, again and again, in the caldron they've all endured. Down to the hard core, he muses, the survivors.

Then come the four Pennsylvania regiments: the 83rd, 91st, 115th, and the 118th. Again small regiments. Pennsylvania men, he smiles his approval, are good to have around: stubborn in fights, and in the camps the first to get shelters built and to find food.

When he reaches the last Pennsylvania regiment, the 118th, he pauses a bit longer.

"Morning, Major Cline."

"General. Great day, sir."

"It is indeed."

Mounted, a bit apart from the 118th's regimental staff, a sergeant major quietly waits, alone, watching him.

Recognizing the lone rider, Chamberlain smiles, "Well, Patrick DeLacy. You're a long way from the 143rd Pennsylvania this morning, but I'm sure there's a good reason for that."

"Yes, sir," DeLacy grins. "Seemed like this was too good to miss. Something I'll be wantin' to tell my young'uns about."

"And, General," he adds, "as this might be our last chance, I'd consider it a great honor if the General would now like a sip from my canteen."

Chamberlain grins. It's something personal between the general and the sergeant. Last summer, on that hot day before Petersburg, just before Chamberlain was wounded, he'd remarked to an aide that he'd give a lot for one sip of water. Sergeant Patrick DeLacy, nearby, heard him and, rolling on his side to free his canteen, offered it to Chamberlain.

"Here you are, General. Try mine."

Chamberlain, his lips burned by the sun and wind and his mouth dry as cotton from anticipating their charge on the Confederate breastworks across the field, would have given a $20 gold piece for a drink of cool water. Instead, he thanked DeLacy but returned the canteen untouched.

"Thanks, Sergeant. I appreciate it. But where we're going you may need it more than I. You keep it. But thanks."

At first DeLacy didn't quite understand but, within an hour, he'd needed that canteen for other wounded soldiers and he began to understand. And, as he came to know Chamberlain better, he understood more. So it seemed that whenever Chamberlain was in a hard fight and the 143rd Pennsylvania was anywhere near, Patrick Delacy somehow fought near the Maine general.

Now Chamberlain grins again, extends his hand, "It would give the General great pleasure."

DeLacy passes his scarred, wooden canteen to Chamberlain, and the general takes a deep drink of the cold water, corks the canteen, and returns it.

Then he extends his hand, "Thank you, Sergeant Major. May you always have cool water."

Turning, Chamberlain retraces his lines, continuing to check his men and to speak with individuals. He takes a commander's pride that they've worked hard to spruce worn uniforms, to scrape Virginia-red clay from long-traveled boots, and to clean rifles and bayonets so they glisten in the morning's sun. It seems to him that the men are standing taller than he's seen them in a long, long time, and he reckons that he's not the only one who didn't sleep much last night.

When he's satisfied that all is ready, he takes his place at the right of his command. Mounted behind him are his staff and the Color Guard with the National flag and the triangular-shaped red, white, and blue flag of the V Corps, flapping gently in the early morning breeze.

Well past sunrise now; nothing to do but wait. It's a cold, gray, cloudy spring day, but the sun's rays, lancing through breaks in the clouds and reflecting from small patches of snow still surviving in the hollows of the hills, have begun to warm and brighten the scene.

Across the valley Chamberlain can see the Confederate camp extending several miles on the far slope. Lee's soldiers there are breaking camp.

Groups of them already have begun to cluster near the road, talking among themselves until their bugles summon them. Others are finishing scant meals, putting out campfires, folding tents or, for the last time as Confederate soldiers, packing personal belongings in their gray blanket rolls. Once the surrender ceremony ends, most of them will begin the long walk home.

"Colonel Spear," he orders, "when this is over send a detail through their camps. Orders or not, a lot of them won't bring us their rifles and bayonets and flags. They'll leave them behind in their camps. Pick them up."

Gray sergeants call out, and the little clusters of men talking beside the road dissolve as soldiers take up their rifles and move to form ranks. Then a dozen bugles repeat the command.

Lee's infantrymen seem in no hurry to get to this day's unpleasant duty, and their sergeants don't press the matter as they would another time, on another field. Let the Bluebellies wait a bit. A small victory, but *something.* Then bugles sound "Assembly", and the scattered groups of men form companies, regiments, brigades, for the last time.

Officers shout commands and the Confederate ranks face down the road, toward Appomattox Court House. Behind him, Chamberlain hears a nervous staff officer exclaim, "Here they come!"

"Here they come!" The call he'd heard on so many battlefields after the artillery stopped, the smoke began to lift, and they could see the long lines of Gray infantry before them.

"Here they come!"; so different in meaning this day. As if anyone needed to be told anyway.

In a long, four-abreast column, the remnants of Lee's Army of Northern Virginia begin their march up the stage road.

The Confederate soldiers were in no hurry to begin, but by the time their first rank reaches the Appomattox, crossing the road below the village, sore muscles limber to their work, and the men instinctively shift into the swinging, loose-gaited stride they've practiced so long. Jackson, fallen at Chancellorsville, called them his "foot cavalry," and so they were, marching more than twenty-five long miles and, when they got where they were going, still able to raise their Rebel yell and charge their enemies' breastworks; convinced that they could beat any Yankee army anywhere, any time.

Not today. Not today. Despite the rations they've been given the last two days, and the rest their sore bodies needed, they're still a half-starved, exhausted, beaten army. They know that, and it's galling. Let one Yank curse, or gesture, or just cheer a little too much for a man to bear, they've decided, and, well, we'll just see what happens.

On they come, warming to the task, approaching the Appomattox and the waiting Blue gauntlet.

Swaying Confederate battle flags, a different arrangement of red, white, and blue than Chamberlain's Stars and Stripes, seem to be everywhere. More flags, it seems, than men; crowning the column with red. The flags survived where men did not.

When they marched up the Cumberland Valley two years ago, Chamberlain reflects, there were more than 70,000 of them. Took pretty near a full day to pass. Not now.

As they come closer, he's sure that he's made the right decisions about the ceremony. Needlessly humiliating their beaten enemy now would be a big mistake. Long ago he'd decided that the Johnny Rebs were fighting for the wrong cause, but they'd fought hard and well, and had earned Billy Yank's respect: one soldier to another. And now Chamberlain would do his part to make them countrymen again.

His soldiers, who'd faced those Gray ranks on so many fields for nearly four years, would honor them. Chamberlain has seen to it.

He didn't clear his plans with anyone. Long ago he'd learned that sometimes it's best not to ask permission in the army. Just ordered that it be done, and never mind the consequences. Grant would agree; he banked on that. And Stanton and those like him are too far away to interfere. Yes, they'd given their enemies their due.

A single rider leads the Confederate column. The Appomattox bridge is down, but the stream is shallow and his horse crosses it in a few dainty steps. A stone's throw behind, more horsemen; then the first infantrymen follow. They too splash across the cold water as if it weren't there.

There's no mistaking the solitary rider. Slender, neatly uniformed, a soldier's bearing, deeply scarred face: Major General John B. Gordon. Gordon, 33 years old, nine times wounded in action.

At Gettysburg, a Confederate prisoner-sergeant (who should have known what he was talking about) told Chamberlain, "You'll know him when you see him, Colonel. Reckon he's the best looking soldier I ever saw." Chamberlain, closely watching Gordon now, reckons the sergeant had a good eye.

Gordon, normally very erect, slumps in his saddle this April morning. His head hangs low, almost on his chest. His eyes seem fixed upon a point in the road directly between his horse's ears. Heavy spirited, dreading the duty at hand, as he nears Chamberlain, Gordon continues to look neither right nor left.

Lee's soldiers are known for singing as they march: "Dixie," "The Bonnie Blue Flag." Not this morning. There is no sound at all, except the muffled clop of the horses' hooves on the sandy road and, behind them, the splashing of many feet at the stream.

When Gordon is nearly abreast of Chamberlain, the Maine general nods and the sharp tones of a Union bugle break the morning's silence. Then the call is taken up by more bugles from the regiments ascending the hill and the staccato beat of drums. Then Union officers are commanding: "Attention!" "Carry Arms!"

The bugles, the drums, the officers' commands, then the sound of thousands of hands' slapping musket stocks as soldiers shift their bayoneted rifles to the marching salute, echoes across the valley.

Echoes again and again from each of the Union regiments in turn, reaching across the valley to bounce off the hills beyond.

Gordon, startled, snaps erect. His horse, for a moment, stops dead in its tracks. Then, recognizing what is happening, the Southern general reins his horse to face Chamberlain.

He looks directly into Chamberlain's eyes, takes in Chamberlain's own salute.

Then he smiles, a tight but grateful smile, and draws his sword. Those nearby clearly hear the soft, metallic "snick" as the steel blade clears its scabbard.

Gently touching his spurs to his horse's flanks so the animal rears and tosses its head, Gordon whips the sword, upward then out and down in a wide, graceful arc to the tip of his boots. Honor answering Honor.

"General Chamberlain. My compliments, sir."

"Thank you, General. Mine to you, sir."

Then, after quietly calling over his shoulder, "Honor for Honor, General Evans," Gordon again turns his horse toward the village, gestures "Forward" with his sword, and continues toward the courthouse.

After Gordon come a half-dozen mounted officers, his staff; then the first of many Confederate flags; and finally Gordon's infantrymen, some still stamping cold Appomattox water from their worn shoes. As they clear the stream and lean into the long hill, they straighten their ranks and their steps again take on a rhythmic pace.

As the Confederate regiments pass, Chamberlain sits quietly on Charlemagne, acknowledging their salutes, scanning their ranks, sharing thoughts with nearby Colonel Ellis Spear or with Sergeant Thomas McDermott.

Their late enemies now march between his soldiers' lines. Gordon's men are terribly thin, with long, shaggy hair and beards. Dirty. Worn. Wearing all shades of butternut, brown, yellow, and gray. Part uniform, part civilian dress picked up somewhere along the way.

Chamberlain smiles at one who wears a ragged gray, forked-tail coat and torn trousers but sports a new hat adorned with a long ostrich feather. He wonders where the soldier found the hat.

Most of the man's companions, however, wear dark, ragged, sweat-stained, dirty, shapeless, slouch hats, also picked up somewhere along the way.

"My God!" he hears an awestruck voice behind him, "no wonder we couldn't kill 'em all. Turn them sidewise, you can't even see them!"

Another quietly agrees, "Poor souls! You'd have to throw a shelter half over them 'afore they'd cast a shadow!"

As each regiment passes, Confederate officers and men return Chamberlain's salute, measuring him with the same mix of curiosity and amazement he feels for them.

"Hard to realize, Ellis," he shares, "that we're this close, but we're not fighting. We've all survived. Hard to grasp; hard to handle."

He identifies most of the passing units by their battle flags. For the rest of his life he'll link himself with those flags and the roads, and fields, and woods the flags have seen.

"There's the regiments that held that sunken road at Antietam, Ellis. That road piled four or five deep with the dead, ours and theirs."

"And that brigade behind it counterattacked Hooker's men in the cornfield. Remember that field? Every stalk cut down by Minie balls, as if by a scythe. Will you ever forget?"

As Stonewall Jackson's old brigade passes, he shares, "Adelbert Ames told me that when we were pinned down before that awful wall outside Fredericksburg Jackson begged Lee to let him turn those regiments loose to finish us off, but Lee wouldn't do it."

"They are the ones who swept away Howard's corps at Chancellorsville, Ellis. Howard didn't have a chance of stopping them; no one could have stopped Jackson that day. And now here they are, or what's left of them. A handful; beaten right into the ground, but still holding their heads up."

"There's the 2nd Virginia, General."

"Yes. Can't be more than seventy of them. I expect they had a thousand when the war began."

"And here come the 4th and 5th Virginia, sir. About fifty men in each regiment. And there's the 27th behind them. Can't have more than twenty men around that flag."

"Jackson's corps. I'm told that they left more than 6,000 men at Gettysburg. I suppose that most of the ones who aren't here now are there or at Sayler's Creek."

Triggered by the ragged, passing column, Chamberlain's memories also pass in review.

There's Cobbs' Georgians; and Chamberlain's eyes narrow, his face hardens in remembrance. Fredericksburg again; that same stone wall below Marye's Heights. Never forget it. Worse even than Gettysburg. Two Union corps bled white before it. Got the 20th Maine through the town. Then we charged, just like all the other regiments before us.

And some of these very men piled us up, just the way we knew they would. We never got near that wall. Then we couldn't go forward or back. Couldn't raise our heads; just had to lay there and take it. Take it for hours. The sharpshooters. And the cold, the awful cold. Almost Christmas time. Never felt so cold before or since. I'll never forget it; don't want to ever forget it.

In the fading December afternoon he'd dragged bodies to a shell hole. Then Colonel Joshua Chamberlain, 20th Maine Regiment, stacked the Union corpses, one atop the other, until he'd built a slight protection from the snipers' constant fire.

He'd never forget the sound of bullets searching for him but striking those bodies.

At last, after dark, he'd led the survivors back across that wide, corpse-strewn field. They brought with them all the wounded they could find; the dead would have to wait.

I'll never forget Fredericksburg, he shakes his head, nor those Georgians behind that stone wall. And here they are, he muses, a few paces away.

He wonders if they recognize his flag as he does theirs.

"There's the 12th Georgia, Ellis," Chamberlain gestures. "Maybe sixty of them in all. Yesterday, General Gordon told me that at Cedar Run, when we were all learning, a captain took command of that regiment after all its other officers had been shot down. After a while he came to General Early and suggested, 'General, my ammunition's near give out. Don't you think we'd better charge 'em.'"

"I think, General," Spear's answers, "that we've all come a long way."

Following the 12th Georgia are McGowan's, Hunter's, and Scales' brigades.

"See 'em, Sergeant Mac? They're the ones who broke our V Corps' lines on White Oak Road. Then we went in and drove them back. Hard to realize that was less than two weeks ago."

Then the last of Anderson's old Confederate Second Corps pass. Not many men or flags; the rest lost at Petersburg or Sayler's Creek.

He recognizes General Henry Heth with A. P. Hill's old corps. Heth in a brand new Confederate uniform; and Chamberlain wonders how he'd been able to manage that.

A. P. Hill: killed just off the Boydton Plank Road. Could it have been only two weeks ago? Hill's corps cost us a lot of good men at Gettysburg: Robinson with 1,200 men down; the Iron Brigade shot to pieces. But it cost Hill too: nearly 3,000 casualties before he realized that Buford and Reynolds meant to hold that ridge.

Numbers. But not just numbers to Chamberlain. Especially with their old enemy so near. Not formless shapes or featureless faces, half-hidden in the smoke and roar of battle, but real men. Just like us. Terribly thin, decimated ranks. Worn, bright-eyed men. The survivors. Just like us.

"There's a lot of staffs without flags, General?" questions Spear.

"Oh, they have the flags, Ellis; just didn't bring them along. They've hidden them or maybe torn them into little pieces and divvied them up among the boys. Let it go; we'd do the same."

The passing column halts, again and again, while regiments before them go through the surrender ceremony. Even with the decimated regiments, there just isn't enough room to accommodate all the surrendered arms, accoutrements, and battle flags at one time. For those who must wait, again and again, it seems a long time.

When Chamberlain's first regiments come to "Carry Arms," the soldiers' marching salute, the Confederate regiments, one after another, straighten their ranks and shift their own muskets to return the Union salute. Then as the Southern soldiers pass between the lines of Northern men, Blue and Gray soldiers eye each other with wary curiosity.

After Chamberlain's initial signal, there are no more bugle calls, no drum rolls, no cannon salutes, no further commands until the marching men near the Mclean house in the village; the soldiers as silent as the hills about them. The only sound their measured pace on the still-hard road.

When John Gordon reaches the McLean house he turns back to take his place behind the first unit to surrender.

The Confederate column halts, closes its ranks. Then, a regiment at a time, Gordon's soldiers shift their rifles from their shoulders to the ground, mindful of their Yankee audience and of their own sergeants' quietly warning, "Do it right. Any of you gives me a shame face here, he'll answer to me later." This time the sergeants mean it; the men know that.

"Left Face!" "Dress Ranks!" Captains make corrections.

Then "Fix Bayonets!" "Stack Arms!"

As the men lay down their rifles, some wish their weapons well.

One Rebel soldier speaks to his as if it were a living person: "Well, wife, I hope I'll never see you again. Reckon you've killed many a Yank; you've done well." Then he pats it a tender "Good-bye," and steps away.

Another simply says, "Good-bye, gun. I'm damned glad to get rid of you. I've been trying to for two long years, and now I've done so."

Of course there are some informal exchanges between Billy Yank and Johnny Reb. No way that couldn't have happened.

A North Carolinian hails: "What regiment is that, Billy?"

"Corn Exchange Regiment, Johnny."

"Well, I recollect we shot the hell outa you folks at the Shepherdstown Ford and at Fredericksburg."

"Yep, Johnny, you did. But I reckon now you're surrenderin' to us."

A Confederate soldier, somewhat recovered from the long march and short rations, sees a beautiful young woman crying on the porch of a home beyond the road and begs his sergeant, "Let me go to her, Les. You see she needs comfortin' real bad."

"Can't do it, Bob. Too many Yanks in between, and you can't break ranks nohow."

For most of them, however, there's little to ease the hard time. Especially when they're told to hang their cartridge boxes on the stacked rifles.

"Fierce hard to give up the last of my cartridges," one mutters to himself. "After all this time, even without the rifle, I just feel naked without my cartridge box."

At last, however, they've placed their weapons, equipment, and ammunition on the piles between Chamberlain's lines, and only their battle flags remain.

Color bearers are ordered to fold them carefully and place them on the stacks of rifles. The flags: torn and faded by sun and wind and time, ripped again and again by shot and shell and rifle fire, often blood-stained. Most bearing the names of the battlefields where they flew: First Manassas, Second Manassas, Fredericksburg, Sharpsburg, Chancellorsville, and Gettysburg. Each name calling forth memories for both armies. Now to be surrendered.

The night before many of the Southern regiments talked it over then voted on it. In Lee's army they do a lot of things that way. Then many of the color bearers tore their flags into little strips and divided them among the men.

"Somethin' to show your young'uns and your grandchildren some day. Now that it 'pears we may have young'uns."

As Chamberlain suspects, some Confederate color bearers have hidden whole flags away; let the Yankees be damned. Many flags, however, still remain, crowning the decimated ranks like red flowers. They must be surrendered. Especially the flags because flags hold men in a far stronger grasp than weapons ever can. Something about a flag.

For many of Lee's soldiers it's very hard to lay down their flags. Tears streak dirty, drawn cheeks. Some men openly sob; then they lay down their flags and turn away. Finally, only the Stars and Stripes remain silhouetted against the Appomattox sky.

Chamberlain quietly waits as one Confederate unit after another lay down their arms, the entire ceremony requiring more than six hours.

During one of the inevitable delays in the procedure, he rides a restless Charlemagne a bit aside and dismounts to ease his own aching hip.

He looks across the valley of the Appomattox. An early April spring, no doubt about that. Bits of snow from the recent squall stubbornly cling to deep crevices of the surrounding hills or nestle in the shaded glens, but even they are rapidly shrinking, seeping into the ground or trickling toward the Appomattox, then to the James and to the open sea below Richmond.

He sees a few crocuses blooming brightly near his boots and marvels at their bright blossoms, grateful for the relief they give his mind, burdened with too many memories this morning. Bright yellows and whites. Curious now, a professor again not a general of infantry, he studies the ground more closely.

Soon he's found more blossoms: pink and white hepatica and white snowdrops. Spring beauties. He picks several, tucking them away in his blouse, something to take to his wife and children, with love.

"Fannie, Fannie," he murmurs to himself, "it's been a long time."

More blossoms nearby: snowdrops, white-flowered poppies he remembers she called "blood roots," and green-hooded jacks-in-the-pulpit. Spring. An end to the old, a beginning anew. For the land, for the men, for the country.

An aide comes for him; the march has begun again.

Another solitary rider nears: Longstreet. Legendary Longstreet. "Old Pete", Lee's war-horse. Bull Run, Gaines Mill, Malvern Hill, Seven Days, Antietam, Gettysburg, and all the battles after Gettysburg. Longstreet, a big man; a slow-moving, stubborn, hard fighter. I expect, Chamberlain muses, it was Longstreet who held out longest against the surrender.

Behind Longstreet he sees Kershaw's men, or what's left of them. Forty percent losses at Antietam and then another forty percent left at Gettysburg after they'd torn through the Peach Orchard and almost took the ridge behind us.

"Remember that wheatfield, Ellis? Six thousand men down. Covered that beautiful golden grass. Bodies everywhere."

"Yes," Spear shakes his head. "Then they came up that hill after us. Runnin' and yellin' like crazy men. I didn't think we had a chance of stopping them."

"And when we did," Chamberlain completes the thought, "we'd left a lot of good men behind too. A lot of good men. A lot of good men gone on both sides."

"There's Pickett's old division," he recognizes several officers, "less than a hundred of them and less than half of them with their rifles."

"Where's Pickett?" Spears asks. "Should be here with them, but he's not. I'm told that he was badly criticized for not being with the ones he left at Gettysburg. Strange."

Hancock, Chamberlain smiles grimly at the thought, told us that we'd done so well on Little Round Top that he was putting us in a quiet part of the line. We'd barely gotten there when they opened with the worst artillery I ever saw. And when it stopped 15,000 men were coming across that mile-wide field.

Pickett had nearly 7,000 men under him that day but, forty-five minutes later, when we'd stopped him cold, most of them were gone. They say he mustered only about 800 men the next morning. And now this is all that's left of them.

When it was all over he'd told Hancock, "General, next time you've a mind to put us in a quiet part of the line, I'd appreciate it if you wouldn't do it."

The last Confederate regiment is pitifully small; smaller than a respectable rifle company. Only the remnants of a regiment. But they cling to their flag, torn and battle-smoked and blood-stained, as if it were dearer than life to them. When their color bearer finally places it on the pile with all the other flags, he bursts into tears and, never mind the consequences, calls to Chamberlain's men.

"Boys, this isn't the first time you've seen that flag. I've carried it on many fields, and I'd rather die than surrender it now."

Someone in blue answers: "Three cheers, boys! Let's have three cheers for Johnny Reb and his flag."

They try, but they can only manage half-hearted, choked cheers before the effort dies. A good many soldiers, Blue and Gray, feel cheeks suddenly wet, voices suddenly dumb. It's that way this April morning.

When it's over Lee's regiments return to their camps to receive their paroles while their former enemies gather up the remaining property of Lee's Army of Northern Virginia.

Lee's infantrymen have emptied their cartridge boxes in the Appomattox street. Studying the piles of cartridges, an enterprising Yankee soldier finds an abandoned cistern, about five or six feet deep, and dumps the cartridges into it. Then he takes live coals from a nearby house, throws them into the hole, and runs. Pretty quick bullets fly in all directions, worse than many of the battles they've seen, and it's a story they'll laugh in telling when they get home.

At dusk, along the crest of the stage road, the lurid flames from still-burning gunpowder flash and sputter and wreathe the blackness with eery many-colored lights. American soldiers always have had a knack for doing things in spectacular ways.

On the hills and in the valley around the Appomattox, campfires again burn. Northern campfires; Southern campfires. Soldiers, not yet leaving the last field on which they'll face each other, finding whatever comfort they can in the flames and in the countless sharing of thoughts about the day.

Chamberlain, the poet part within him never completely overcome by the warrior part, quietly stands again in the gathering darkness, looking across the valley.

He'll never forget the day or the marching men of both sides, for the soldier part always will be closest to his heart. Whether Blue or Gray, soldiers are a part of him, bonded to him by ties that nothing on earth can sever. Ties that those who've not seen battle will never quite understand. For Chamberlain more men had passed by this day than others saw. All the fallen ones had been there too, as if it had been resurrection morning.

Glad as he is to be going home soon now, there are things that he knows he'll miss. But, even as he thinks that, he recalls Longstreet's sharing Lee's reflection after the terrible day at Fredericksburg.

"It's well that war is so terrible, General. Otherwise, we would grow too fond of it."

Lee was right, he's decided. It's time now, or soon will be, for all of us to lay down this part of our lives; to tuck it away in our hearts and minds and only take it out occasionally to recall in private. Then to tuck it away again. Time to begin anew.

LEE

Lee is not with his regiments when they lay down their arms and flags. Not physically present anyway. He meant to be but, over his headquarters' campfire the night before, Gordon and Longstreet persuaded him not to take part.

"You've done your part, General," Gordon had argued. "Drunk the bitter cup. Grant isn't here; left yesterday. He'd not expect you to be present either. Besides, Mrs. Lee's waiting in Richmond. She's not well; needs you. We can do this."

Longstreet's argument had been more persuasive.

"General, you mind the times the boys didn't want you getting any closer to things? They'd take Traveller's bridle and say 'Lee to the rear! Lee to the rear!' They're saying that now, General. Don't want to see you before the Yankee army that way. Listen to them. Ride out of here in the morning, erect and proud, the way they've always known you. They deserve that memory."

He'd smiled. Pete, his old war-horse, so slow yet somehow always able to get to the heart of things. Then he nodded acceptance, reached out to take both their hands.

Morning now. He watches his soldiers forming on the road below, glances about his camp. His last headquarters in the field has been in a grove of white oaks, chestnuts, and Virginia pine; on a hill above the road and the headwaters of the Appomattox. Under other circumstances, a pleasant place.

His headquarters, normally small and simple anyway, now reduced to a campfire, a picket line for the horses, captured tents still bearing a distinct "U.S." on their sides, the captured wagon-ambulance he uses as an office.

After he watches his regiments begin their march to the courthouse, he finishes his breakfast of Union rations: hardtack, fried pork, coffee without milk. Then, wearing a neat, plain gray uniform, he receives his commanders (Gordon and his officers had come earlier because they'd be first in the surrendering column). As always, he is bright, cheerful, thoughtful of each of them, thanking them for past service, encouraging them about their futures. All, however, leave in tears.

About ten o'clock he looks about very carefully, as if memorizing the scene. Then he turns to Marshall and quietly orders, "Strike the tent, Colonel."

Lieutenant Samuel Lovell, with Sergeant William B. Arnold and fifteen men of the 4th Massachusetts Cavalry Regiment, will escort them beyond the Union lines.

They begin quietly, but once the little caravan (a dozen Confederate officers and men; Lee's office-ambulance; a crude wagon bearing all their personal goods beneath a nondescript quilt; another ambulance bearing a badly wounded officer; and the Union cavalrymen) reaches the road, the word quickly spreads.

Confederate soldiers hurry for one last look as Lee passes. Many cheer; some can only raise ragged hats in salute. Many of them reach to shake Lee's hand, murmuring their affection. Time and again, until he's cleared the last Confederate picket line, the Rebel yell they've shared on so many battlefields rings across the valley. Then it fades away. Dimmer, dimmer, gone.

As Lee's party rides toward Richmond the road still is strewn with traces of a phantom army: broken wagons, limbers, and caissons; dead horses and mules; soldiers' knapsacks, bullet pouches, and canteens.

About a dozen miles beyond Appomattox Court House they crest a long hill. Beyond it the road descends into a valley then rises again to disappear on the far horizon. Lee signals a halt then turns to the Massachusetts' officer.

"Lieutenant Lovell, you and your men have come far enough. You'll be needed back with your regiment. I'm within my own country and among friends; I needn't trouble you further. Rejoin your regiment."

He exchanges handshakes with the young officer, wishes him well and a safe journey home.

Then he returns the young officer's salute, just as he'd been taught to do on the Plain at West Point so many years before, waves the wagons forward, and smiles, "Come, gentlemen, let's go home."

Afterword

THE WAR

Following the laying down of their arms and battle flags, the Army of Northern Virginia, like the last patches of snow on the warming fields about Appomattox Court House, simply melt away. Gray soldiers, singly, in pairs, or by squads, go home unmolested, hoping to pick up their shattered lives.

Meanwhile, Grant's armies retrace their steps to the Petersburg-Richmond area, marching in torrents of cold rain. Mud everywhere, and most of the men are without tents. An Irish soldier, Matthew McElroy, unable to sleep while his Down Easter messmate snores through the rain, observes, "A Yankee would sleep if the divil sat on his head."

Scant rations; more long, cold, tiring marches. At night still posting their pickets, as required by regulations, against a phantom enemy. It will take awhile to get used to peace.

Several weeks later they'll march in triumph, along with the "bummers" of Sherman's Western army, down Pennsylvania Avenue in Washington, all led by Major General Joshua Lawrence Chamberlain.

As they return to Richmond-Petersburg, however, their march is interrupted by horrible news.

Good Friday, April 14, 1865, and President and Mrs. Lincoln have invited General and Mrs. Grant to join them for the evening's performance at Ford's Theater in Washington. Grant begs off, pleading earlier plans for a trip to the Northeast. Lincoln graciously accepts the explanation, knowing that at times Mary Lincoln is not pleasant company and that Grant has earned time with his own family. As for himself, Lincoln feels happier than he has in a long, long time. The war clearly is ending; the Union all but totally restored. His plans for healing the terrible rift between the North and the South are firm in his mind. Complete the task of reuniting the country, he's decided; then home to Springfield; perhaps a little lawyering.

During the performance Lincoln's personal bodyguard leaves his post unguarded long enough for a twenty-six-year-old Southern sympathizer-actor named John Wilkes Booth to slip into the presidential box and fire a single bullet into Lincoln's head. The President, carried to a nearby home, lingers through the night, but at 7:22 a.m. the next morning, Stanton announces, "Now he belongs to the ages."

Andrew Johnson, a much lesser man, is sworn in as the President; and the assassin's flight is intercepted on the Garrett farm below Washington. Federal cavalry Sergeant Boston Corbett kills Booth. The other conspirators are given a swift trial then hanged.

Meanwhile, Jefferson Davis has moved what remains of his government on wheels to Greensboro, North Carolina where, on April 12, he confers with Confederate General Joseph E. Johnston.

Davis, still full of fight, urges Johnston to attack Sherman then Grant.

Johnston hears him out then, sadly shaking his head, replies, "Sir, our people are tired of this war, feel themselves whipped, and will not fight."

When Davis' few remaining cabinet members agree with Johnston, Davis adjourns the meeting that they might reconsider. That evening, however, a tired, muddy courier brings Lee's final report. It's official now: the Army of Northern Virginia is gone; and Davis weeps uncontrollably.

He agrees to write Sherman requesting a general ceasefire to allow "Civil authorities to enter into the needful arrangements to terminate the existing war."

When Sherman and Johnston meet, Sherman offers Johnston terms as generous as those Grant offered Lee at Appomattox. Reduced to writing, however, Sherman's proposal is much broader than Grant's, getting into civil and political matters.

He sends a copy of it to Washington for approval, and Edwin A. Stanton, through President Andrew Johnson, rejects it, reprimanding Sherman's "letting them up easy."

On April 26, having no other choice, Confederate General Johnston accepts Sherman's revised terms for the Confederacy, surrendering the 30,000 soldiers of his Army of the Tennessee and almost 90,000 more Southern soldiers scattered throughout North Carolina, South Carolina, Georgia, and Florida.

Meanwhile, Davis, threatened by Sherman on the south, Grant on the north, and Federal cavalry patrols on all sides, has fled Greensboro.

At dawn, May 10, a Wisconsin cavalry regiment surrounds Davis' camp near Irwinville, Georgia. Shackled, Davis is taken to Savannah for the voyage to Fort Monroe, an old Federal post at the mouth of the James River below Williamsburg.

Arriving there on May 23 Davis is immediately hustled into a small, cheerless cell in Casemate #2 of the old fort. There, bound with heavy leg irons and forced to endure the humiliation of having a Federal guard in his cell at all times, he begins a period of solitary confinement that will last for two years. Soon after it begins, to avoid a public outcry, the shackles are removed; the constant guard is not.

At first Davis is accused of conspiring in the plot to assassinate Lincoln. When Federal investigators can't substantiate the charge, it's amended to simple treason. When they also can't prove the lesser charge, Davis finally is released, unrepentant, to go to Canada.

On May 4, 1865 General Richard Taylor surrenders his 42,000 Confederate soldiers in Alabama, Mississippi, and Louisiana. Confederate Brigadier General Samuel Jones then surrenders his Florida command; and, three weeks later Confederate Brigadier General Kirby Smith surrenders the nearly 18,000 men of his Trans-Mississippi Department.

The Confederate army at last is gone, and Union President Andrew Johnson proclaims armed resistance at an end.

Confederate fighting men don't all simply fade away. On May 12 the last battle of the war is fought at Palmito Ranch, beside the Rio Grande River in Texas. A pointless engagement in which the Yankee soldiers seem sure to win but then are driven from the field. Men have been killed and wounded on each side.

Not too far from Palmito Ranch Confederate Brigadier General Joe Shelby, still refusing to surrender his little brigade, gathers his men about him.

"Boys," he tells them, "the war is over, and you can go home. I for one, however, will not go home. Across the Rio Grande there lies Mexico. Who will follow me there?"

Several hundred Confederate veterans will, and they'll pick up another hundred on their ride to the river. Crossing the Rio Grande one quiet June morning they pause long enough to weigh their old Confederate battle flag with stones then sink it into the dark waters. Then they spash across into Mexico, offering their services to the emperor Maximilian.

On the high seas the Confederate cruiser *Shenandoah*, its captain, James Waddell, not knowing that the war is over, continues to raid Federal shipping. When he learns that the fighting has ended, he turns his ship away, skirts Cape Horn, and eludes Yankee warships.

On November 6, 1865, the *Shenandoah* sails into the harbor at Liverpool, England, the Confederacy's Stars and Bars proudly flying from its mast. There its crew will surrender to English authorities.

Perhaps the last of the holdouts, three Confederate scouts sent into the Dismal Swamp below Norfolk in early 1863 to spy on Federal activities, will surrender in July 1866.

UNION BLUE

In 1869 General Ulysses Simpson Grant will replace Johnson as President. Grant will fill his cabinet and other government positions with men he knew in the army; some of them won't serve him well.

During his second term in office gigantic frauds surface. Government positions have been sold; appointments made without regard for the fitness of their occupants; gifts loosely accepted. It's never proved that Grant has been involved, but he's tarred with the same brush as his subordinates.

After his presidency, he lives in a New York City home donated by his admirers and attempts to start a brokerage firm which quickly fails. His friends have his name restored to the army's retired list so he'll have some income to provide for his family.

Dying of cancer, Grant will spend his last days writing his autobiography which Mark Twain will publish then pay him $450,000, allowing Grant, having met his obligations, to die in peace on July 1, 1885.

Brigadier General John Rawlins will briefly serve as Grant's Secretary of War before dying of tuberculosis in March 1869.

Colonel Horace Porter also will serve Grant's administration, personally avoiding the scandals that wreck Grant's second term in office.

Adam Badeau will become Grant's personal historian.

General William Tecumseh ("Cump") Sherman follows Grant in commanding the army. Frustrated by Sheridan's prolonged campaign to subdue the Indian Nations in America's West, he'll retire in 1884 and die in New York City in 1891.

Confederate General Joseph E. Johnston marches bareheaded in Sherman's long funeral procession then dies of pneumonia within five weeks.

After the war General Philip H. Sheridan is instrumental in forcing the government of Napoleon III to withdraw its military support of the Mexican emperor Maximilian, easing that threat to the United States.

Sheridan then will serve in a number of army posts, including command of the Division of the Missouri, as his cavalry fight the Plains Indians. In 1884, following the retirement of Sherman, he becomes the commanding general of the army.

Sheridan dies on August 5, 1888 and is buried in Arlington National Cemetery, the site of Mary Custis Lee's ancestral home.

General George Gordon Meade also remains in the army after the war, serving in a number of insignificant military departments and continuing to assert that he'd not been given proper credit for his role at Gettysburg and the subsequent campaigns under Grant's command. He is commanding the army's Division of the Atlantic in Philadelphia when he dies of pneumonia in November 1872.

General Gouverneur Kemble Warren, the chance of further advancement in the army destroyed when Sheridan relieved him of his command of the V Army Corps at the Battle of Five Forks, spends his remaining military service in the army's Corps of Engineers, writing prolifically on engineering matters and credited with numerous engineering projects.

Meanwhile, he fights to have a Board of Inquiry review his relief by Sheridan at Five Forks. After much delay and a prolonged hearing, that board not only exonerates him completely but also criticizes the manner of his relief. Warren dies at his home in Newport, Rhode Island, on August 8, 1892.

Colonel Ely Parker, Seneca chieftain and Grant's personal secretary, becomes the U.S. Commissioner of Indian Affairs in 1869. He uses that office to try to improve the lot of the American Indians but is brought before the House of Representatives on trumped-up charges of malfeasance of office. He's found innocent but resigns and retires to civilian life.

General Joshua Lawrence Chamberlain, six times wounded in action, many times cited for bravery, awarded the Congressional Medal of Honor, chosen to lead the last Grand Review of the army in Washington, returns to Maine to resume his teaching post at Bowdoin College.

He's bored with his classroom although, before he's done, he'll teach every class the college offers except one in mathematics.

He finds it difficult to adjust to civilian life until he's asked to run for governor of Maine; then he wins by the largest majority in the state's history and is returned to the office three times.

In 1876 he's elected president of Bowdoin College and attempts to modernize its curriculum by offering more courses in science, deemphasizing religion, and introducing an ROTC program.

Chamberlain remains very active in Union veterans' affairs and is popular as a speaker and writer throughout the North and the South. He often returns to Gettysburg, the last time in 1913 just before the Grand Reunion of Union and Confederate veterans there.

Joshua Lawrence Chamberlain dies in June 1914, of complications from the wound he'd received at Petersburg fifty years before, and is buried in a quiet little New England cemetery just off the Bowdoin College campus.

General George Armstrong Custer, perhaps second only to Confederate General Jeb Stuart in his flamboyance but second to none in his battlefield aggressiveness, will continue his tempestuous military career after the war. Reduced, by the peacetime shrinking of the army, from major general to lieutenant colonel and from commanding a division of cavalry to Executive Officer of the 7th U.S. Cavalry Regiment, he finds the adjustment very difficult.

He is acclaimed as Sheridan's finest Indian-fighting general, accolades he welcomes, but he also is accused of massacring a sleeping Indian village on the Washita River in dead winter and in 1867 is court-martialed for leaving his command in the field to visit with his wife.

On June 26, 1876, Custer will die with 266 of his officers and men after charging a large Sioux camp on the Little Big Horn River in Montana.

CONFEDERATE GRAY

General Robert Edward Lee, following the surrender of his army at Appomattox Court House, returns to Richmond to his invalid wife, Mary Ann Randolph Custis, and to his family. He refuses many offers of lucrative positions to accept the presidency of Washington College, a small, destitute college (now Washington and Lee University) in Lexington, Virginia.

Like Jefferson Davis', Lee's United States' citizenship is revoked but, unlike Davis, he never is imprisoned. Also unlike Davis, Lee tries (unsuccessfully) to have his citizenship restored. Lee's and Davis' citizenship finally will be restored during President Jimmy Carter's administration.

Lee's prestige and his influence among Confederate veterans do much to bring about a peace and reconciliation that many thought not possible. In 1870 he'll die of the heart disease that first troubled him during the Gettysburg campaign, perhaps the most respected general in American military history.

Traveller, Lee's war-horse, accompanies his master to the campus of Washington College where, in the late afternoons when Lee's work is done, the general scoops up little boys and girls for brief rides on the famous Traveller. He'll survive his master by a few months and now rests beneath a soldier's headstone just outside the little, white campus chapel; nearby should Lee need him again.

General James Longstreet will settle in Georgia after the war. There he'll make two great mistakes which will cloud the remainder of his life: after Lee's death he blames Lee for the Confederate defeat at Gettysburg; and he becomes a Republican.

For years Longstreet will not be invited to reunions of Confederate veterans. He comes anyway, stubborn to the end. At the last one he attends, he limps slowly down the aisle, all alone, in the uniform of a lieutenant general of the Confederate States Army. Before he reaches the front aisle, his old comrades are all on their feet, giving him a standing ovation.

He dies in 1904 at the age of 83. His younger widow will survive, eventually serving in the early 1940s (she in her 80s) in another American war fashioning Air Corps' bombers on an assembly line in Marietta, Georgia.

General John Brown Gordon, who began his military career with the Confederacy as the Company Commander of the "Racoon Roughs" and ends it as one of Lee's two corps commanders, returns to Atlanta, Georgia after the war.

For the next forty years he'll be the idol of much of the South: a great public speaker and writer, mostly known for his reminiscences of the Confederate soldier.

Gordon serves in the U.S. Senate (1873–1880, 1891–1897) and as Georgia's governor, 1886–1890. Like his Northern counterpart, Joshua Chamberlain, he is a prime mover in Confederate veterans' affairs, serving as the first commander of the United Confederate Veterans from 1890 until his death on January 9, 1904.

General George Edward Pickett, following Appomattox, continues to brood over the loss of his division at Gettysburg. Long after the war he unexpectedly comes with Confederate Brigadier General John Singleton Mosby to visit Lee, a meeting which Mosby later described as "singularly cold"; afterward, Pickett bitterly tells Mosby, "That man destroyed my division."

Colonel Charles Marshall, aide and confidant to Lee, will practice law in Baltimore and research Lee's command.

Major Edward Boykin, taken prisoner at Appomattox Court House, returns to his native South Carolina where he resumes the life he'd known before the war and, like many other Blue and Gray veterans, writes his memoirs of his service to the Confederacy.

Confederate nurse Phoebe Yates Pember, her patients reassigned soon after the fall of Richmond, finds herself alone in deserted Chimborazo Hospital. Having no idea where she will go or what she will do, she opens a little box to tally her more important possessions: a small amount of now worthless Confederate currency and a silver Federal ten-cent piece. She trades her entire fortune for five coconut cakes and a box of matches. Then she leaves the hospital for the last time.

Taking a small apartment in Richmond, penniless, alone, frightened, she finds that somehow word of who she is and what she did for wounded Confederate soldiers for four years gets around. Then each day, until she returns to her native Charleston, South Carolina, she receives a steady stream of visitors, offering food, work, whatever she needs, just as she'd tried to answer the needs of her patients.

Doctor John Claiborne and Drummer Boy John L. G. Woods also survive Appomattox to return to their homes and rebuild their lives.

Selected Bibliography

Bearss, Edward, and Christopher Calkins. *The Battle of Five Forks.* Lynchburg, Va.: H. E. Howard, Inc., 1985.

Bernard, George S., Ed. *War Talks of Confederate Veterans.* Petersburg, Va.: Fenn and Owen, Publishers, 1892.

Boykin, Edward M. *The Falling Flag.* New York: E. J. Hale and Son, 1874.

Calkins, Christopher. *From Petersburg to Appomattox, April 2–9, 1865.* Farmville, Va.: The Farmville Herald, 1990.

———. *The Battles of Appomattox Station and Appomattox Court House: April 8-9, 1865.* Lynchburg, Va.: H. E. Howard, Inc., 1987.

———. *The Final Bivouac: The Surrender Parade at Appomattox and the Disbanding of the Armies.* Lynchburg, Va.: H. E. Howard, Inc., 1988.

———. *Thirty-Six Hours Before Appomattox: The Battles of Sayler's Creek, High Bridge, Farmville and Cumberland Church.* Farmville, Va.: The Farmville Herald, 1980.

Chamberlain, Joshua Lawrence. *The Passing of the Armies.* Dayton, Ohio: Morningside Press, 1991.

Claiborne, Dr. John H. *Seventy-Five Years in Old Virginia.* New York: Neale Publishing Company, 1904.

Davis, Burke. *To Appomattox: Nine April Days.* New York: Rinehart and Company, 1959.

———. *The Long Surrender.* New York: Random House, 1985.

Freeman, Douglas Southall. *Lee's Lieutenants.* Vol.3. New York, Charles Scribner's Sons, 1944.

Gerrish, Theodore. *Army Life: A Private's Reminiscences of the Civil War.* Portland, Me.: Hoyt, Fogg and Donham, 1882.

Gordon, John B. *Reminiscences of the Civil War*. New York: Charles Scribner's Sons, 1903.

Grant, Ulysses S. *The Personal Memoirs of U.S. Grant*. Vol 2. New York: Charles L. Webster and Company, 1886.

Johnson, Robert U., and Clarence C. Buel, Eds. *Battles and Leaders of the Civil War*. Vol. 4. New York: Thomas Yoseloff, Inc., 1956.

Longstreet, James. *From Manassas to Appomattox: Memoirs of the Civil War*. Philadelphia: J. B. Lippincott Co., 1896.

Merington, Marguerite, Ed. *The Custer Story*. New York: Devin-Adair Company, 1950.

Pember, Phoebe Yates. *A Southern Woman's Story*. New York: G. W. Carleton and Company, 1879.

Porter, Horace. *Campaigning with Grant*. New York: The Century Company, 1897.

Pullen, John J. *The 20th Maine*. Dayton, Ohio: Morningside Press, 1984.

Rodick, Burleigh Cushing. *Appomattox: The Last Campaign*. Gaithersburg, Md.: Olde Soldier Books, Reprinted 1987.

Schaff, Morris. *The Sunset of the Confederacy*. Boston: John W. Luce and Company, 1912.

Sheridan, Philip H. *Personal Memoirs of P. H. Sheridan*. Vol. 2. New York: Charles L. Webster and Company, 1888.

Stern, Philip Van Doren. *An End to Valor*. Boston: Houghton Mifflin, 1958.

Stiles, Robert. *Four Years Under Marse Robert*. Washington, 1903.

Taylor, Walter H. *Four Years with General Lee*. New York: Reprinted by Bonanza Books, 1962.

Trudeau, Noah A. *The Last Citadel: Petersburg, Virginia, June 1864–April 1865*. Boston: Little, Brown and Company, 1994.

———. *Out of the Storm: The End of the Civil War, April–June 1865*. New York: Little, Brown and Company, 1994.

Trulock, Alice Rains. *In the Hands of Providence: Joshua Chamberlain and the American Civil War*. Chapel Hill, N.C.: University of North Carolina Press, 1992.

Urwin, Gregory J. *Custer Victorious*. Nebraska: University of Nebraska Press, 1983.

Wheeler, Richard. *Witness to Appomattox*. New York: Harper and Row, 1989.

Woods, John L. G. Article in *Confederate Veteran* (1919).

Woodward, C. Vann, Ed. *Mary Chesnut's Civil War*. New Haven: Yale University Press, 1981.